OCEANOGRAPHY

by
GILBERT L. VOSS

Illustrated by
SY BARLOWE

GOLDEN PRESS • NEW YORK

WESTERN PUBLISHING COMPANY, INC.

RACINE, WISCONSIN

FOREWORD

Today as never before man is turning to the sea for his livelihood and perhaps for his ultimate survival. The present world population is about 3½ billion people. In the next 30 years, it will become 6 billion, and in 70 years, if the present growth rate continues, it will be 12 billion. These people will need food and water. With less space available on land for agriculture and ever greater demands for fresh water, man must turn to the sea for much of his sustenance. This book surveys what the sea can be expected to yield and how it can be harvested. We must learn even now how to manage these tremendous resources without over-exploiting and how to control and reduce pollution in the sea.

Many people have helped with this book. I particularly wish to thank my colleagues from the Rosenstiel School of Marine and Atmospheric Science, University of Miami, especially the late Dr. F. Koczy who helped plan it, Dr. and Mrs. Donald Moore who read all of the text, Dr. Eugene Corcoran who read and commented on the chemistry section, Dr. Walter Düing who read and criticized the physical section, Drs. Frederick Bayer and Jon Staiger who helped with technical matters on animals, Dr. Lloyd Burkle for supplying research information to the artist, and my wife for her review of the text and her encouragement. Any errors or faults are my own responsibility.

G. L. V.

CONTENTS

3

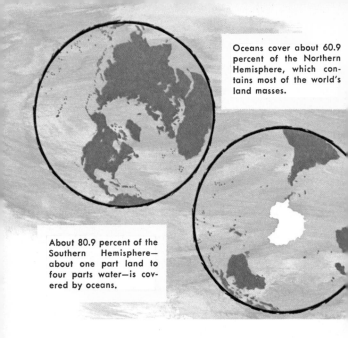

Oceans cover about 60.9 percent of the Northern Hemisphere, which contains most of the world's land masses.

About 80.9 percent of the Southern Hemisphere—about one part land to four parts water—is covered by oceans.

OCEANS AND OCEANOGRAPHY

Seas cover about 70.8 percent of the earth. They are unevenly distributed. In the Southern Hemisphere, the oceans circle the globe. Much of the world's weather originates over or near the Antarctic ice-cap, an area of heavy storms and gales.

Oceanographers generally recognize three major oceans: Atlantic, Pacific, and Indian. Where they meet is commonly known as the Southern or Antarctic Ocean. All oceans are separated mainly by land masses except the Southern Ocean. It surrounds

Antarctica and is defined on its northern boundaries by the Subtropical Convergence at about 40° south latitude. The Arctic Ocean is surrounded on all sides by land, with the entrances protected by shallow sills. Altogether, the oceans and the seas cover slightly more than 140 million square miles, or 361 million square kilometers.

AREA, VOLUME, AND AVERAGE DEPTH OF THE WORLD'S OCEANS AND SEAS

BODY	AREA (in square kilometers)	VOLUME (in cubic kilometers)	AV. DEPTH (in meters)
ALL OCEANS (excluding seas)	321,130,000	1,322,198,000	4,117
Atlantic Ocean	82,441,000	323,613,000	3,926
Pacific Ocean	165,246,000	707,555,000	4,282
Indian Ocean	73,443,000	291,030,000	3,963
LARGE MEDITERRANEAN SEAS	29,518,000	40,664,000	1,378
Arctic Mediterranean	14,090,000	16,980,000	1,205
American Mediterranean	4,319,000	9,573,000	2,216
Mediterranean Sea and Black Sea	2,966,000	4,238,000	1,429
Asiatic Mediterranean	8,143,000	9,873,000	1,212
SMALL MEDITERRANEAN SEAS	2,331,000	402,000	172
Baltic Sea	422,000	23,000	55
Hudson Bay	1,232,000	158,000	128
Red Sea	438,000	215,000	491
Persian Gulf	239,000	6,000	25
ALL MEDITERRANEAN SEAS	31,849,000	41,066,000	1,289
MARGINAL SEAS	8,079,000	7,059,000	874
North Sea	575,000	54,000	94
English Channel	75,000	4,000	54
Irish Sea	103,000	6,000	60
Gulf of St. Lawrence	238,000	30,000	127
Andaman Sea	798,000	694,000	870
Bering Sea	2,268,000	3,259,000	1,437
Okhotsk Sea	1,528,000	1,279,000	838
Japan Sea	1,008,000	1,361,000	1,350
East China Sea	1,249,000	235,000	188
Gulf of California	162,000	132,000	813
Bass Strait	75,000	5,000	70
ALL ADJACENT SEAS	39,928,000	48,125,000	1,205
ATLANTIC AND ADJACENT SEAS	106,463,000	354,679,000	3,332
PACIFIC AND ADJACENT SEAS	179,679,000	723,699,000	4,028
INDIAN AND ADJACENT SEAS	74,917,000	291,945,000	3,897
ALL OCEANS AND SEAS	361,059,000	1,370,323,000	3,795

From The Oceans by H. U. Sverdrup, M. W. Johnson, and R. H. Fleming, Prentice-Hall, Inc., New York, N.Y. 1942.

THE OCEAN DEPTHS were a complete mystery to early mariners, and many scientists thought that the ocean bottom was a level "abyssal plain." Magellan, on his voyage across the Pacific in 1521, attempted the first deep-sea soundings by lowering a cannonball overside on the end of a 400-fathom (730-meter) line. Of course, he failed to touch bottom. By soundings made first with lead on a piano wire and now by electronic echo-sounders, the depths and general features of the ocean floors are mapped. Much more information is needed to fill in the details, but it is known that in some places the ocean floors are very rugged, with great mountain chains, canyons, plains, and trenches.

Bathymetric map of Atlantic Ocean, showing bottom features

THE AVERAGE DEPTH OF THE OCEANS is five times the average height of the land, and in general the continents stand nearly 3 miles (4.8 kilometers) above the ocean floor. Indeed, over half of the area of the globe exceeds a depth below sea level of 1¼ miles (2,000 meters), while there are more than 50 known ocean depths greater than 3¾ miles (6,000 meters). In the deepest sounding ever recorded, made in 1962 by H.M.S. Cook in the Mindanao Trench east of the Philippine Islands, the echosounder registered 11,515 meters, or over 7 miles. If Mount Everest (29,141 feet or 8,760 meters), the world's highest mountain, were dropped into the Mindanao Trench, it would be covered by water to a depth of about 1.6 miles.

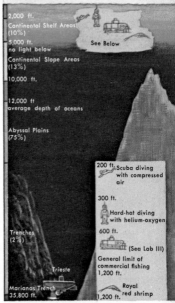

Shown below are the percentages of various types of sediments at given depths in the oceans.

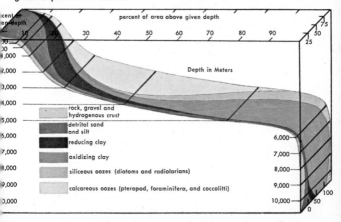

OCEANOGRAPHY is a modern science with ancient roots. The study of the seas actually began with the early voyages of discovery. As the seas were mapped, some information was gathered also about ocean currents and sea life. Systematic studies were begun when scientists were added to the crews, but most of the early scientists had special interests in mind rather than the study of the oceans generally.

Oceanographic studies were made by such men as Benjamin Franklin, who first plotted the course of the Gulf Stream; Edward Forbes (1815-1854), who studied the geographical and vertical distribution of marine plants and animals; and Matthew Fontaine Maury, whose *The Physical Geography of the Sea* (published in 1855) is ranked as the first

THE GULF STREAM was first mapped by Benjamin Franklin while he was U.S. Postmaster General. Whalers had long known the mighty current's course and speed, but merchant captains persisted in trying to sail against it. Franklin hoped that use of his chart would help ships to avoid the current and to speed the mail in crossing from Great Britain to America. His map is reproduced below.

H.M.S. CHALLENGER, equipped with sail and steam, sailed from England in December, 1872, on a voyage to investigate the physics, chemistry and biology of the world's oceans. She returned 3½ years later, having sailed 68,890 miles. The results of this voyage occupy 50 large volumes and form the basis of the present science of oceanography.

textbook of oceanography. The first modern oceanographic cruise was the historic expedition of H.M.S. *Challenger*, with Wyville Thomson as chief scientist. Nearly every major nation soon sent out vessels to study the seas.

Shore laboratories were built to study the data and specimens collected. The first were the biological laboratories and fisheries stations, such as the Zoological Laboratory at Naples and the fisheries laboratory at Woods Hole. Oceanographic institutions established solely to study ocean phenomena came much later. Today, only a few large, highly organized institutions, mostly government supported, conduct oceanographic research in its broad sense. Special aspects of marine biology are studied at small laboratories along most coasts.

VOYAGES to study the seas started with Capt. Cook's cruises to the Pacific in 1768-79. Since that time, many ships have been sent on voyages of exploration, biological and chemical research, and meteorological investigations. The most historic are listed in the chart below.

MAJOR OCEANOGRAPHIC VOYAGES

Ship	Scientific Leader	Nation
Paramour Pink	Halley	Great Britain
Endeavour	Cook	Great Britain
Erebus & Terror	James Ross	Great Britain
U.S. Expl. Exped. Peacock & Vincennes	Wilkes	United States
Lightning	Carpenter & Thomson	Great Britain
Porcupine	Carpenter & Jeffrys	Great Britain
Challenger	Thomson	Great Britain
Tuscarora	Belknap	United States
Blake	A. Agassiz	United States
Albatross	A. Agassiz	United States
Hirondelle	Prince of Monaco	Monaco
Princesse Alice	Prince of Monaco	Monaco
Princesse Alice II	Prince of Monaco	Monaco
Hirondelle II	Prince of Monaco	Monaco
Vitiaz	Makaroff	Russia
Fram	Nansen	Norway
Belgica	de Gerlache	Belgium
Valdivia	Chun	Germany
Siboga	Weber	Holland
Discovery	Scott	Great Britain
Gauss	von Drygalski	Germany
Antarctic	Nordenskiold	Sweden
Scotia	Bruce	Scotland
Albatross	A. Agassiz	United States
Francais	Charcot	France
Porquoi Pas?	Charcot	France
Deutschland	von Drygalski	Germany
Michael Sars	Hjort	Great Britain & Norway
Dana	Schmidt	Denmark
Meteor	Haber	Germany
Discovery	Kemp	Great Britain
Discovery II	Kemp	Great Britain
Willebrord Snellius	van Riel	Holland
Albatross	Pettersson	Sweden
Anton Bruun	Ryther	United States
Glomar Challenger	Bader and Gerard	United States

Missing from this list is that important group of permanent research ships kept at sea the year around by major marine nations of the world. Most of them are converted commercial ships or private yachts, though ships specially designed for research are replacing them.

FROM HALLEY TO 1969

Geographic Area	Field of Study	Date
Atl. and Southern Oceans	physical, magnetic variation	1699-1700
South Seas	physical, biological	1768-1777
Southern Ocean	physical, biological	1839-1843
Pacific	bottom depths, biology	1838-1842
North Atlantic	general, biology	1868
North Atlantic	general, biology	1869-1870
World Oceans	general, biology	1872-1876
Pacific	deepsea temperatures, sediments	1875
Florida & Caribbean	general	1877-1880
Pacific	general	1899-1900
North Atlantic	general	1885-1892
North Atlantic	general	1892-1897
North Atlantic	general	1898-1911
North Atlantic	general	1911-1914
World Wide	general	1886-1889
Arctic Ocean	general	1893-1896
Antarctic	general	1897-1899
Atlantic, Indian and Southern Oceans	general	1898-1899
Dutch E. Indies	general	1899-1900
Antarctic	general	1901-1903
Antarctic	general	1901-1903
Antarctic	general	1901
Antarctic	general	1902-1904
Eastern Pacific	biology	1904
Antarctic	physical, biological	1903-1905
Antarctic	physical, biological	1908-1910
Antarctic, Atlantic	physical, biological	1911
North Atlantic	general	1910
World Wide	general, biological	1920-22, 28-30
South Atlantic	chemistry	1925-1929
Antarctic	general	1925-1926
Antarctic	general	1930-
Dutch E. Indies	general	1929-1930
World Wide	physical, geological	1947-1948
Indian Ocean	biology	1963-1965
Atlantic	geology	1969

MODERN OCEANOGRAPHIC SHIPS may carry a crew of 30 or more scientists and stay at sea for several months. Smaller vessels carry only half a dozen scientists and stay at sea only a few days.

Oceanographic ships are costly because of their special research equipment. Radar, loran, deka and satellite navigational equipment can fix positions to within a few hundred yards or closer. Precision depth recorders give depths down to 30,000 feet (9,000 meters) and more with great accuracy. Large winches capable of holding 50,000 feet (15,000 meters) or more of wire cable are used for trawling and coring in great depths. Weight alone would break ordinary wire cable in these lengths, so that specially engineered cable or tapered cable is used. Smaller high-speed winches with lighter wire are used to collect plankton or for hydrographic work.

THE ALBATROSS, commissioned in 1882 by the U.S. Fish Commission, was the first ship designed and built especially for oceanographic research. During her career, she covered more sea miles and had more deep hauls made from her decks than from any other oceanographic vessel. Several fisheries vessels of the United States have carried her proud name.

THE DISCOVERER, commissioned in 1967, was constructed specifically for research in Atlantic waters. The *Discoverer* is capable of 150 days at sea and has a sustained speed of 16 knots. It is equipped with complete laboratory facilities. The *Oceanographer*, a sister ship, is stationed in the Pacific.

Most large ships have special powder magazines where high explosives are stored for underwater explosion studies. Sounds made by animals or by man are listened to and recorded in seismic and acoustic studies. An oceanographic vessel must also have dry and wet laboratories for studying specimens, analyzing water samples, and similar work. The scientists need libraries for quick reference and electronic computers for analysis of data.

To keep from ripping delicate nets apart or breaking expensive cables and losing trawls, an oceanographic ship must be able to go very slowly—from 0 to 4 knots. Usually this is accomplished with variable-pitch propellers or with diesel-electric or steam turbine propulsion, allowing complete control of speed. To maintain position while "on station," a rotating propeller or a fixed "bow thruster" is often mounted in the bow of the ship.

THE OCEANOGRAPHIC LABORATORY must be large and efficient to provide for the special and diverse needs of the teams of scientists. Roughly 15 scientists are needed ashore for every scientist at sea, or 15 days of shore work for every scientist's day at sea. Large ships are more self-sufficient.

The large oceanographic laboratories ashore must be equipped with such specialized instruments as

MAJOR OCEANOGRAPHIC INSTITUTIONS

Woods Hole Oceanographic Institution	Woods Hole, Massachusetts
Lamont-Doherty Geological Observatory	Palisades, New York
Rosenstiel School of Marine and Atmospheric Science, University of Miami	Miami, Florida
Scripps Institution of Oceanography	La Jolla, California
Department of Oceanography, University of Washington	Seattle, Washington
National Institute of Oceanography	Wormley, England
Musée Océanographique	Monaco
Institute of Oceanography	Halifax, Nova Scotia
Danmarks Fiskeri-og Havundersøgelser	Charlottenlund Slot, Denmark
Institut Océanographique	Paris, France
Institut für Meereskunde	Kiel, Germany
New Zealand Oceanographic Institute	Wellington, N. Z.
Oceanographic Laboratory	Edinburgh, Scotland
Instituto Español de Oceanografía Marina	Madrid, Spain
Oceanografiska Institutet	Gothenburg, Sweden
Institute Okeanologii	Moscow, U.S.S.R.
Department of Oceanography, Oregon State University	Corvallis, Oregon

INSTITUTIONAL RESEARCH SHIPS

Akademik Kurchatov	U.S.S.R.	Institute of Oceanology
Mikhail Lomonosov	U.S.S.R.	Institute of Oceanology
Discovery	Gr. Britain	National Institute of Oceanography
Atlantis II	U.S.A.	Woods Hole Oceanographic Institution
Vema	U.S.A.	Lamont-Doherty Geological Observatory
Thomas Washington	U.S.A.	Scripps Institution of Oceanography
James N. Gillis	U.S.A.	Rosenstiel School of Marine and Atmospheric Science, Miami
Thomas G. Thompson	U.S.A.	University of Washington
Discoverer	U.S.A.	NOAA* Oceanographic Laboratory
Walter Herwig	W. Germany	Institut für Meereskunde
Eltanin	U.S.A.	National Science Foundation
Yaquina	U.S.A.	Department of Oceanography, Oregon State University

*National Oceanic and Atmospheric Administration, Dept. of Commerce.

electronic computers, analytical laboratories with special spectrographic instruments, carbon 14 and rubidium-thorium dating labs, culture rooms for microbiology, and other facilities. Machine shops, electronic shops, net rooms for trawls, carpenter shops, and marine divisions for ship maintenance and operation are also needed for a self-sufficient oceanographic laboratory.

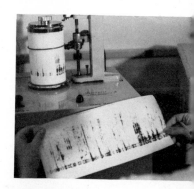

FISH SOUNDS, recorded from the ocean bottom by tape recorder attached to an underwater pickup, are being analyzed by this scientist in a laboratory ashore. The sounds are being played back and recorded on a sound spectrograph to produce a picture for visual study.

SCRIPPS INSTITUTION OF OCEANOGRAPHY at La Jolla, California, is the world's largest oceanographic laboratory.

▼

SPECIAL SHIPS FOR SEA STUDIES have been developed. The U. S. Navy now owns Jacques Piccard's bathyscaphe *Trieste* but seldom uses it. Numerous other submersibles, such as the *Aluminaut, Benjamin Franklin,* and *Deep Star,* are now being used to make observations and to collect samples in the deep sea.

FLIP, an unusual vessel useful for studying currents and waves, is towed horizontally to its station. It is then flipped into a vertical position by water ballast. It rides deep in the water, providing a stable platform from which the scientists can work.

THE TRIESTE descended to the bottom of the Marianas Trench (35,800 ft.) in January, 1960. The trip took 5 hours. The 50-foot vessel was weighted by 10 tons of iron held by electromagnets. To ascend—a 3-hour trip—the iron was released. The ship rose because her tanks were filled with gasoline, which is lighter than water.

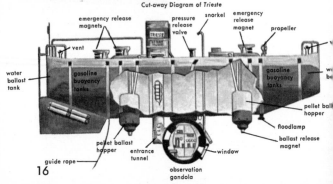

Cut-away Diagram of *Trieste*

emergency release magnets

vent

pressure release valve

snorkel

emergency release magnet

propeller

water ballast tank

gasoline buoyancy tanks

gasoline buoyancy tanks

w... be...

pellet ballast hopper

floodlamp

pellet ballast hopper

entrance tunnel

window

ballast release magnet

guide rope

observation gondola

FLIP—Floating Instrument Platform—is a 335-foot vessel that is towed to its working area. When after ballast tanks are flooded, ship upends into a vertical position with only 55 feet remaining above surface. This provides scientists with stable work platform.

THE ALUMINAUT, a 51-foot submarine, carries a crew of three down to 15,000 feet. This depth allows exploration of about half the ocean floor.

DEEPSTAR uses mercury ballasts to achieve a quick descent, going down tail first. To rise, weight is dropped. Vessel comes up nose first.

Scientist is lowering equipment into the sea from the chains or "hero's platform" that juts out at the side of the ship. The platform may plunge under waves in heavy weather.

EQUIPMENT used in oceanography today is greatly advanced over the simple surface thermometer, dipnet, and cannonball sounding leads of early investigators. Surface and mid-water currents are now measured with radio beacons, Swallow buoys, and current meters (p. 41). Temperature, salinity, and other water characteristics are determined electronically. Sampling the bottom for sediments, rocks, and bottom life is done with corers, grabs, dredges, and trawls. Precision echo-sounders give bottom profiles, and deep-sea cameras view the bottom with accuracy. Television cameras sweep the bottom in shoal depths. Closing water bottles are used to collect water samples for chemical analysis and microbiological samplings in mid-depths. Large and active animals are collected with trawls at high speeds.

Seismographs investigate the earth's crust for thickness. Heat probes measure the earth's temperature; gravimeters, the variations in gravity; and magnetometers, the amount of magnetism. Light penetration is measured with photometers, and bioluminescence counters measure the light given off by living organisms.

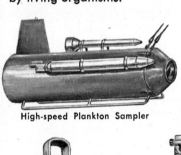

High-speed Plankton Sampler

Underwater Television Camera

Nansen Bottle

STD (Salinity-Temperature-Depth)

MARINE GEOLOGY

Geological oceanography is concerned with the formation of the ocean basins, the origin of sea water, bottom topography, the origin of bottom sediments, and interactions between the sea and the shore. Formerly much of marine geological research was conducted by land geologists, using material brought into the laboratory, but today's marine geologists are bona fide oceanographers.

Cameras and television have been important instruments in recent studies of ocean-bottom topography, and now the *Aluminaut*, *Deep Star*, *Benjamin Franklin*, and other manned underwater vessels are contributing greatly to marine geology. With these vessels (pp. 16-17), geologists cruise along the ocean floor, viewing the bottom and taking bottom grabs, short cores, and rock samples as desired.

DEEP-SEA SOUNDINGS were made successfully by Sir James Ross in 1840, but systematic soundings to construct profiles of the ocean bottoms were first gathered by M. F. Maury (p. 8). A bathymetric chart of the Atlantic Ocean was produced by him in 1854.

As late as 1912, only about 5,969 soundings had been made over the world in depths of 1,000 fathoms (1,800 meters) or greater. Each sounding was made by wire or rope and took several hours. Today, with echo-sounding instruments, soundings are made continuously and transcribed onto graph paper, showing bottom-contour details at a glance.

THE FIRST successful deep-sea sounding was made by Sir James Ross in the South Atlantic on January 3, 1840, from H. M. S. *Erebus*. The line was lowered from a small boat. Two other pulling boats were attached to the boat lowering the line so that alternate pulling, as needed, would keep the reel directly over the sounding line to get an accurate reading.

THE ECHO-SOUNDER is used today to make continuous deep-sea soundings. An electronic signal is sent down from the hull of the ship, and its return echo is recorded on moving graph paper, shown at right. Because of pressure and temperature differences, which affect the speed of transmission, corrections must be applied to the readings. These are derived from published tables.

21

Deep-sea cores are kept under refrigeration.

THE FIRST DEEP-SEA DEPOSITS collected in a systematic fashion were taken by Sir John Murray aboard H.M.S. *Challenger*. Thousands of samples have since been collected from ocean floors around the world. For study of sediment distribution, short cores only a foot or so long are sufficient, but to study the earth's past history and ocean formation, cores several hundred feet long are needed. Getting one of these long cores to the surface requires special and expensive apparatus, plus great skill in seamanship and gear handling.

Cores are refrigerated at about 41° F. (5° C.) to prevent chemical changes from taking place. Sections of the core are dated by carbon 14 or other methods to determine their age. By selecting sections of the core containing the skeletons of tiny calcareous animals known to live above the thermocline (p. 46), the region where the temperature drops rapidly, and by analyzing their oxygen isotope relationship by means of the mass spectrograph, the exact temperature of the water in which the animals lived can be determined. From these studies, temperature changes and when they occurred are known with accuracy over millions of years.

Marine geologists also examine the rocks on the sea floor and in the strata beneath. Rock samples can be obtained with strong steel dredges, but the strata sequences can be determined directly only by ocean drilling. By use of the seismograph, much has been discovered about depths of sediment layers, thickness of the crust, and the earth's core. Another instrument used now is an electronic device called a "sparker" that sends strong signals through the bottom sediments. Two ships some distance apart may also be utilized for these studies. One drops depth charges; the other records the discharge and the resulting sound waves as they travel down through the soft sediments and rock layers of different density, then return to the surface.

STUDY OF FOSSILS in deep-sea cores has revealed much about the kinds of life that swarmed in ancient seas, the conditions under which they lived, and the geological age of the major groups of animals. Spores of land plants found in deep-sea deposits have helped unravel the story of both land and sea.

At top left, an oceanographer looks through the "file" of refrigerated cores. At right, he examines a section of one of these cores.

THE ORIGIN OF OCEAN BASINS has been explained by various theories over the years. A generally accepted theory explaining the origin of the solar system says it began as a great whirling cloud, or disc, of cosmic dust and gases. The nebulous mass was probably cold at first but became hot as it contracted. The sun formed at the disc's nucleus, where the heat was highest. At varying distances, tails from the cloud concentrated to form the planets. Our earth was probably first a liquid or molten mass.

Today, apparently confirmed by the discoveries made aboard the *Glomar Challenger* (pp. 38-39), Wegener's theory of continental drift is back in vogue. He proposed that the continents were formerly united. Stresses within the earth broke the continents apart, and they drifted into their present positions, floating upon the molten magma deep beneath them. Rock cores taken by the *Glomar Challenger* from the mid-Atlantic ridge provide evi-

dence that the sea floor is spreading outward, pushing the continents apart. Forces involved are apparently convection currents in the earth's molten interior. Mid-oceanic ridges today indicate areas of sea-floor spreading.

Apparently, during the Cretaceous Period the present continents consisted of a single giant land mass set in the middle of a truly world ocean. Within this megacontinent fractures appeared, delimiting great crustal plates that slowly drifted away from each other. Antarctica swung southward to its present polar position; India moved northward to crunch into the Asian continent, forming the Himalayas at their point of impact. North America slowly drifted away from Europe, forming the North Atlantic in its wake. Probably as late as 100 million years ago, South America began its drift away from Africa. The Pacific is the oldest ocean; the South Atlantic is the youngest.

THE ORIGIN OF OCEAN WATERS was originally believed to be from great rains that formed from water vapors in the early atmosphere. Scientists now say that there was originally only enough water vapor in the atmosphere to fill the ocean basins to about 16 percent of their present volume. It is more likely that the water came from rocks during the earth's formation. A considerable percentage of water is bound up in rocks, especially in the silicates but also in such halogens as the chlorides and fluorides. As the earth's crust hardened and contracted, these waters were released. Over the still-warm surface of the earth, the water undoubtedly formed clouds, then condensed and fell as rain. Later, as the earth cooled, water spouted up from volcanoes, hot springs, and other outlets. These processes have slowly filled the ocean basins and even overrun the land.

In composition, the original seas were remarkably like those of today, but over the millions of years, salts have been added as rain combined with atmospheric carbon dioxide to form carbonic acid that chemically weathered the rocks.

Even from these sources, only about half of the volume of the water in the oceans can be accounted for in rocks from the crust. Perhaps the remainder came from the upper part of the earth's mantle.

SHORE LINES vary from stable rocky headlands to ever-changing sandy beaches. Generally, the larger the size of the particles forming a beach, the steeper its slope. Shores reflect the sea's restlessness, its tremendous constructive and destructive forces. Often they indicate an uplift or a slow settling.

BEACH PROFILE

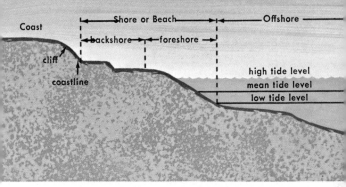

Coast

cliff

coastline

Shore or Beach

backshore · foreshore

Offshore

high tide level
mean tide level
low tide level

ROCKY SHORES are usually steep. They erode continually but much more slowly than sandy shores, which are unstable and may change from storm to storm. Sandy shores build up as the waves come from one direction, then wash away as the waves change direction. Usually sand is deposited on the up-current side of a cape and carried away from the down-current side.

Since mud consists of very fine silt particles that are carried easily even by weak currents, muddy shores are characteristic of protected beaches. They are not steep and may be very extensive. Other types of beaches are cobble, shingle, boulder, and gravel, named for the size of the material of which they are made.

If erosion is heavy, especially in built-up areas, sea walls, jetties, and groins are construct-ed to protect shores. They should be shaped so that they not only protect the beach but also dampen the effect of the waves. If this is not done, erosion may be heavy on the down-current side of the wall.

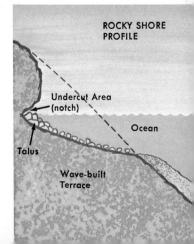

ROCKY SHORE
PROFILE

Undercut Area
(notch)

Ocean

Talus

Wave-built
Terrace

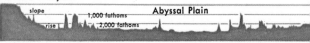

OCEAN BOTTOMS are divided into three broad categories and a number of subdivisions: continental margins (shelf, slope, and rise), ocean basin floor (floor, plains, hills, and rises), and mid-oceanic ridges. All of these can be seen above in the profile of the Atlantic from Martha's Vineyard to Gibraltar.

OCEAN-BOTTOM TOPOGRAPHY

OCEAN-BOTTOM TOPOGRAPHY has been known only recently. Originally the bottom of the seas was pictured as a vast plain—smooth, featureless, and probably frozen and devoid of all life. This was the "abyssal plain" theory (p. 6). But with the deep-sea soundings made by Sir James Ross and M. F. Maury (p. 21) came the realization that ocean bottoms have features much the same as those on land. Not until the electronic echo-sounder came into use was the true picture of the great ocean basins seen. Now we know that the floor of the oceans resembles the land, with plains, terraces, canyons, hills, volcanoes, and mountain peaks.

Today the general features of ocean bottoms throughout the world are known. Many details will not be filled in for years of course. Scientists at the University of Miami, for example, only recently released charts of the detailed features of the bottom of the Straits of Florida, so important to the marine life of the Gulf Stream.

OCEAN DEPTHS exceed 20,000 feet (6,000 meters) in about 50 areas, mostly in the Pacific. The type of map on the opposite page is a Hammer-Aitoff projection, developed by Soviet scientists. It shows the water areas with the least distortion, hence is now used widely in oceanography. The land masses are greatly distorted, however, in this type of map.

Mid-Atlantic Ridge

Azores

Gibraltar

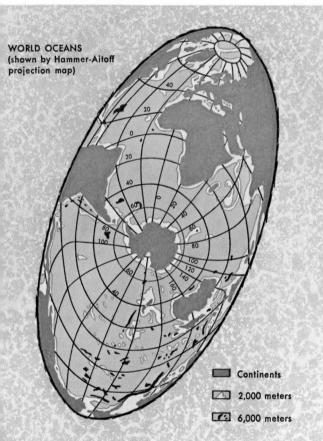

WORLD OCEANS
(shown by Hammer-Aitoff
projection map)

Continents

2,000 meters

6,000 meters

THE CONTINENTAL SHELF, the shallow platform bordering most continents and large islands, extends from the shore to a depth of about 600 feet, or 180 meters. Actually, the average outer limit is 433 feet (132 meters), but this varies widely because of the variety of ways in which the shelf is formed. Once it was thought to be formed only by wave erosion, for waves were believed effective down to 600 feet (180 meters). Now the average limit of wave erosion is known to be about 65 feet (20 meters) or less.

In some places the continental shelf is up to 600 miles (965 kilometers) wide; in others it is narrow or completely missing, as along newly formed coast lines of steep mountain ranges. Much of the shelf is smooth and featureless, but in places it is cut through by valleys and trenches, and by the beginnings of submarine canyons (p. 35).

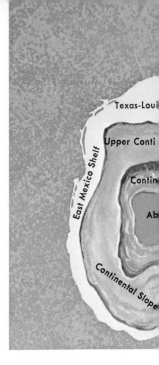

THE CONTINENTAL SLOPE extends from the edge of the continental shelf to a depth of about 6,600 feet (2,000 meters), where the ocean floor begins. The slope is not steep, averaging 3° to 4° on mountainous coasts and only 2° on shallow coasts. Slopes may be smooth, terraced, or even hilly. Submarine canyons also cut through the slope. The shelf and slope area together constitute only about 16 percent of the ocean bottom.

THE CONTINENTAL RISE is the gradual slope at the base of the continental slope, at a depth of about 6,600 feet (2,000 meters). Generally smooth, it is bounded seaward by abyssal plain and hills.

THE OCEANIC BASIN FLOOR, by the geologist's definition, extends from the continental rise to the other side of the ocean. Oceanographers, in contrast, class a basin as the deep area on each side of a ridge.

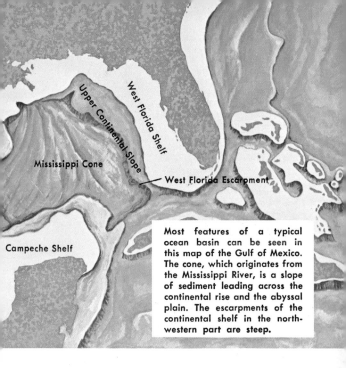

Most features of a typical ocean basin can be seen in this map of the Gulf of Mexico. The cone, which originates from the Mississippi River, is a slope of sediment leading across the continental rise and the abyssal plain. The escarpments of the continental shelf in the northwestern part are steep.

ABYSSAL PLAINS are the most characteristic feature of the ocean floor. First identified in 1947 when the continuous-recording, deep-sea echo-sounders came into use, they are flat plains with only slight irregularities, and they occur at the bottom of each of the ocean's major depressions and also in marginal seas. Often associated with them are abyssal hills that rise above the plain for a few meters to as much as several hundred meters and may be from a few hundred meters to a few kilometers wide. Abyssal hills are thought to be formed of deposits of fine silts from the shelf area. The plains also have fan-shaped deltas of fine sediments from the continental areas.

OCEAN RISES rise directly from the deep ocean floor without connection to the continental margins. The Bermuda Islands, for example, are located on the Bermuda Rise.

THE MID-OCEANIC RIDGE, the world's grandest mountain system, extends for 40,000 miles (64,000 kilometers) through the Norwegian Sea, the North and South Atlantic, the Indian and South Pacific, and the Arctic oceans. The ranges of this undersea mountain system divide the major oceans and cause them to be different on either side. Other ridges, not part of the main system, may subdivide the basins into smaller areas. The general height of the Mid-oceanic Ridge is 1½ to 2 miles (2 to 3 kilometers); its width is about 620 miles (1,000 kilometers).

Lagoon

Patch Reef

Lagoon Floor

FRINGING REEF STAGE

OCEANIC ISLANDS, typically small, rise abruptly from the ocean floor, independent of any continental mass. Unless part of a mountain chain, they are isolated from each other. St. Helena, Ascension, and the Azores rise from the same mountain chain. The Marquesas, Samoa, Tahiti, and the Hawaiian Islands are isolated.

SEAMOUNTS are volcanic mountains that have not reached the surface. Over 1,200 steep-sided seamounts exceeding half a mile (1 kilometer) in height have been described. Because they have basaltic bedrock, they are known to be of volcanic origin.

Guyots are flat-topped seamounts. The tops are often about 5,000 feet (1,500 meters) below present sea level. About 200 are now known. Guyots are considered to be former seamounts that have been worn down by wave erosion and changing ocean levels. The Pratt Guyot shown here is located in the North Pacific.

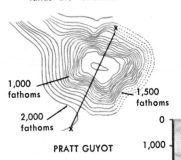

1,000 fathoms

1,500 fathoms

2,000 fathoms

PRATT GUYOT

0

1,000

2,000 fathoms

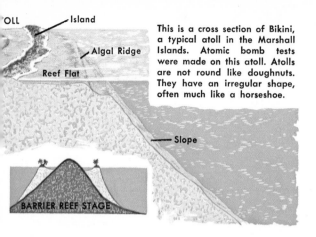

Island

ATOLL

Algal Ridge

Reef Flat

Slope

BARRIER REEF STAGE

This is a cross section of Bikini, a typical atoll in the Marshall Islands. Atomic bomb tests were made on this atoll. Atolls are not round like doughnuts. They have an irregular shape, often much like a horseshoe.

ATOLLS are coral reefs, roughly circular and often projecting above the sea's surface. Commonly they consist of a ring of islands surrounding a shallow lagoon. All of the land present is of coral origin.

Darwin proposed the first acceptable theory of how atolls were formed. He said they started when corals grew around the edge of an island, forming a fringing reef. Then the land began to sink slowly, but the corals, growing rapidly, were able to build up a coral platform and maintain themselves in near-optimum conditions near the surface. The land continued to submerge, the corals surviving but always doing best on the outer side. Overcrowding and lack of sufficient food and sunlight gradually retarded the growth of corals inside the ring.

Eventually only a small part of the island remained, surrounded by a lagoon and an offshore, or barrier, reef. In time the land sank beneath the sea, leaving only the ring of coral. This was now an atoll.

It has been suggested that all atolls were formed much in this manner, either by the sinking of the bottom or the rising of the sea level. According to the latter theory, the level of the sea rose due to melting of ice in interglacial periods, such as we are now experiencing. This would by itself or in conjunction with the sinking of the island create an atoll. Echo-tracings support this explanation of atoll formation.

Sometimes a sudden change in the ocean level has "drowned" a reef, sinking the corals below the level at which they can grow.

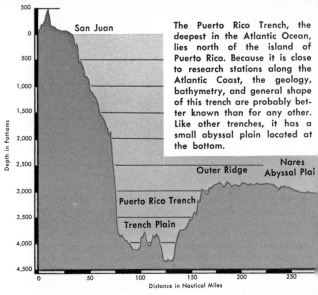

The Puerto Rico Trench, the deepest in the Atlantic Ocean, lies north of the island of Puerto Rico. Because it is close to research stations along the Atlantic Coast, the geology, bathymetry, and general shape of this trench are probably better known than for any other. Like other trenches, it has a small abyssal plain located at the bottom.

TRENCHES, or Deeps, exceed 19,700 feet in depth (6,000 meters). The deepest is the Marianas Trench (p. 6). Trenches occur in all major oceans. Usually they are named either for the ship or for the oceanographer who discovered them. A few, such as the Wharton Deep in the Indian Ocean and the Byrd Deep in the South Pacific Basin, occur in the middle of the abyssal plains and near the center of the ocean basins. Most are found near to and paralleling coasts, island arcs, or submarine ridges.

Trenches may be narrow but almost invariably are flat across the bottom. In cross section, a trench is V-shaped.

One theory of how the trenches formed suggests that the bottom was infolded by a movement of the mantle and crust toward each other. The fold is downward due to the weight of the water above. According to another and more recent theory, the trenches formed as cracks that were torn in the crust and then partly filled by magma from below.

Echoes from depth recorders bounce back and forth along the sides of trenches and make it difficult to identify the bottom echo. Also, the signal is weakened in travel through the great depth of water—5 to 6 miles (8 to 10 kilometers).

SUBMARINE CANYONS are widely distributed along the edges of continents. Two well-known canyons in United States waters are the Hudson Canyon off New York and the Monterey Canyon off California. Submarine canyons have all the characteristics of land canyons, including branching side canyons, steep cliffs, and narrow, winding gorges. Usually they occur off mouths of rivers and bays, sometimes in front of land canyons or below flat deltas.

Cables laid across these canyons are often broken by landslides. Currents created by these slides sometimes attain considerable velocity as they sweep down the canyon. These are called turbidity currents. Heavy with silt, they have a strong scouring and erosive power. Many geologists now believe that these turbidity currents are the principal cutting forces that cause submarine canyons, though the hardness of some of the rock walls in the canyons seems to make this doubtful. Others believe the canyons were cut by rivers from the land.

Shown below are two branches of Scripps Canyon. Turbidity currents of high velocity are needed to cut such steep walls. Sand slides can be seen at several places.

The two canyon profiles below —that of Monterey Canyon and of Grand Canyon—indicate that the canyons were probably formed by similar processes of erosion.

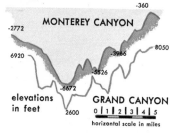

MONTEREY CANYON

-360
-2772
6920
-3966
8050
-3526
-6672
2600

elevations in feet

GRAND CANYON

0 | 1 | 2 | 3 | 4 | 5
horizontal scale in miles

MARINE SEDIMENTS consist of the remains of organisms and of a variety of minerals and salts that accumulate on the ocean bottom. Their composition varies with depth and nearness to land. There are three basic kinds: (1) shallow-water near-shore, (2) shelf, and (3) deep-sea. The study of marine sediments forms a separate division of marine geology.

SHALLOW-WATER NEAR-SHORE SEDIMENTS are mainly inorganic. They come primarily from the land and are carried to the sea by erosion, runoff water, and streams. Many sediments are moved miles from their point of origin, carried far out into the sea by currents before being deposited. Heavier, coarser sediments are dropped first, usually in rather shallow water. Lightweight, fine sediments are carried by the currents into deep water.

SHELF SEDIMENTS, transported by currents and by wave action, are found on open coast lines from the beach out to the edge of the continental shelf. They consist of the calcareous (lime) remains of organisms (biogenic); of salts, greensand, or glauconite (an iron-potassium silicate), and other substances precipitated from sea water and deposited in place (authigenic); and of materials eroded by waves from rocks, older sediments, and detrital matter.

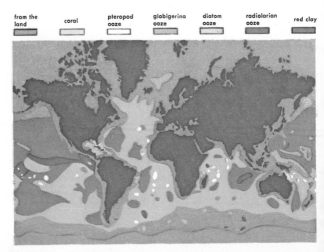

from the land | coral | pteropod ooze | globigerina ooze | diatom ooze | radiolarian ooze | red clay

DEEP-SEA SEDIMENTS cover nearly two thirds of the earth's surface. Most are pelagic in origin—that is, from the upper layers of the open ocean. They are formed of the skeletons of planktonic (floating) plants and animals, of cosmic and wind-blown dusts and clay. Near land masses, terrigenous (land) sediments such as blue, green, and red muds are carried to the depths. Pelagic sediments are divided into two general categories: oozes and red clay.

Oozes are subdivided into four kinds. Diatom ooze consists largely of the siliceous skeletons of diatoms (tiny plants) and occurs primarily in polar waters. Globigerina ooze, found in temperate and tropical seas, is composed of skeletons of several kinds of foraminifera, tiny one-celled animals. Pteropod oozes, found also in temperate and tropical seas but less widespread, are derived from the skeletons of snails that live in the open sea. In deep parts of the Pacific and Indian oceans, the predominate ooze consists of the silica shells of radiolarians.

Below about 16,000 feet (4,800 meters), in the oceans' greatest depths, even silica is largely lacking. The bottom is formed mostly of red clay sediments derived from volcanic and other dusts from the land, ionic dust, and other fine particles. Hard objects such as manganese nodules (of meteoric origin) and sharks' teeth or the ear bones of whales are found in the clay. By radiocarbon dating of cores, it has been determined that pelagic (offshore) sediments are deposited at a rate of from .025 of an inch (0.01 cm.) to over 24 inches (60 cm.) per 1,000 years; and red clays at from 0.1 of an inch (.07 cm.) to 0.5 of an inch (0.2 cm.) per 1,000 years.

TYPES OF SEDIMENTS

Diatom Ooze

Radiolarian Ooze

Pteropod Ooze

Globigerina Ooze

Red Clay

THE GLOMAR CHALLENGER is the world's first deep-ocean drilling ship (shown at right). Drill shafts and diamond cutter heads are lowered section by section to the sea floor in the drill casings shown at left stacked on the ship's foredeck. Art at extreme right shows the ship drill casings, and bottom depth approximately to scale. Special propulsion equipment, plus bottom sonar beacon, maintain her position.

MOHOLE, LOCO, AND JOIDES are names (or abbreviations) of programs for sampling the earth's surface beneath the seas by drilling.

The first project was Mohole, conceived by a group of geologists at Scripps Institution of Oceanography. They proposed to drill through the earth's outer crust and into the mantle from offshore, where beneath the sea the earth's crust and hence the depth that must be drilled to reach the mantle is much less than on land. The Jugoslav seismologist Andrija Mohorovicic had discovered that seismic waves are bounced off the interface where the crust and the mantle meet, hence this interface was named for him—the Mohorovicic Discontinuity, shortened to Moho. The plan to drill into the earth's mantle in the sea was called Mohole.

Some scientists felt that more immediate scientific and economic returns could be realized by studying the history of the deep sedimentary layers of the ocean floor. In 1962, a program called LOCO (from Long Core) was suggested by scientists at the Rosenstiel School of Marine and Atmospheric Science, the Lamont Geological Observatory, the Woods Hole Oceano-

graphic Institution, and the Scripps Institution of Oceanography. Before this proposal died, Miami scientists aboard the *Submarex* retrieved a core 56.4 meters (about 175 ft.) in length.

In 1964, the same group of institutions formed JOIDES (Joint Oceanographic Institutions Deep Earth Sampling). Based on a committee's survey, a number of drill sites were selected along the Atlantic Coast. The first successful drillings were accomplished by the *Coldrill* on the Blake Plateau in depths from 25 to 1,030 meters; the core lengths were 120 to 320 meters. Supported by the National Science Foundation in 1969, the *Glomar Challenger*, a specially designed and constructed deep-ocean drilling ship, has successfully drilled in great ocean depths. The results of this program far exceeded expectations, forming the basis for new theories concerning continental drift, sea-floor spreading, crustal origins, and mineral resources in the sea.

about 3 miles deep

PHYSICAL OCEANOGRAPHY

Physical oceanographers study temperature gradients, density differences, cause and structure of waves, tidal movements, sea levels, currents, light penetration, the relationship between the sea and the air, and similar features. They are concerned primarily with the causes of these phenomena, hence this branch of oceanography is largely theoretical. Information from nearly all other branches of oceanography is needed to understand these factors, for they influence and are influenced by all other elements of the seas.

A physical oceanographer is usually highly trained in physics and mathematics. He spends much of his effort in testing and proving theories, but the oceans are his laboratory, test tube, and home. His basic tool is the oceanographic ship (p. 12), but he also utilizes specialized equipment.

REVERSING THERMOMETERS are used to measure temperature distribution in the sea. A series of reversing thermometers fastened to a hydrographic cable is lowered to the desired depth. With a trigger mechanism, the thermometers can be inverted to record the temperature against pressure.

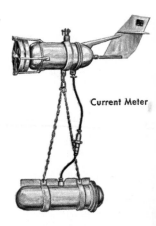

Current Meter

The BT (bathythermograph), can be lowered and retrieved rapidly. A tracing on a gold-coated slide records the temperature on a depth scale.

The STD (salinity/temperature/depth) sensor records salinity and temperature continuously as a function of depth.

A PRECISION DEPTH RECORDER (PDR), a very accurate type of echo-sounder, is used to determine the topography and depth of the ocean bottom. By careful calculations that depend on the temperature and other factors, echo-sounder readings are converted into depths, then bottom contours.

CURRENT speeds are measured with moored current meters, radio buoys, current crosses, and other means. The accuracy of the measurement depends on the preciseness of the navigational fixes by the ship's officer on the bridge. Deep currents are found and traced by such devices as Swallow buoys drifting with the current beneath the surface and tracked by ships above.

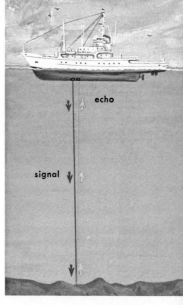

echo

signal

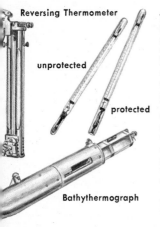

Reversing Thermometer

unprotected

protected

Bathythermograph

THE GEK (geomagnetic electrokinetograph) measures electromagnetic potential differences, which are known to be affected by the movement of ocean currents. A GEK consists of two electrodes towed some distance apart behind a moving vessel. The potential difference between them is measured. Such measurements can also be obtained between two stationary electrodes attached to a telegraph cable. For example, the volume of water passing through the Straits of Florida between Key West and Havana is determined rather accurately in this way.

41

AVERAGE LIGHT PENETRATION IN SEA

650 ft. (lower limit of most plant life)

2,300 ft. (no visible sunlight below this level)

Deep Sea (light only from bioluminescent organisms)

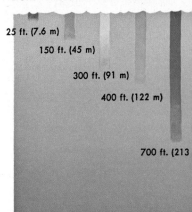

LIGHT PENETRATION IN BERMUDA WATER

25 ft. (7.6 m)

150 ft. (45 m)

300 ft. (91 m)

400 ft. (122 m)

700 ft. (213

LIGHT from the sun is the source of the heat that warms the seas. In turn, the heat storage of the seas plays an important role in the earth's climate. Light is required also for photosynthesis, the process by which plants manufacture food. The tiny floating plants (phytoplankton) of the open sea account for an estimated 80 percent of the total photosynthesis on earth.

Sea water is a barrier to light. Usually not enough light penetrates beyond 1,600 feet (490 meters) to be visible to the human eye, except in the clearest tropical oceanic water. In coastal waters, light may not penetrate more than 160 feet (50 meters); in very turbid waters, probably not more than a few inches or centimeters. Minute particles of dust, organic debris, and water molecules absorbs and scatter the light rays.

Photographic plates exposed in the mid-Atlantic have shown that in the clearest water a few red rays may penetrate to 328 feet (100 meters). At 1,640 feet (500 meters), blue rays were still plentiful, but green rays were absent. Even at 3,280 feet (1,000 meters), a slight darkening of the plates occurred, but at 5,560 feet (1,700 meters) no light was present.

Clear waters of the open sea appear blue because blue rays are absorbed less and are scattered more, even by the water molecules themselves. Thus the blue of the oceans is comparable to the blue of the sky. Coastal waters appear greenish because the larger number of particles in suspension reflect back most of the colors. Yellow pigments abundant in some coastal waters absorb the blue and violet light, and the combination results in the sea's green or yellowish-green cast.

Florida peninsula photographed from Apollo 9 spacecraft. Dark blue of the Gulf Stream can be seen off Straits of Florida.

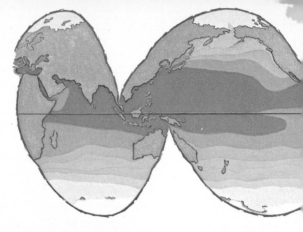

TEMPERATURES of ocean waters vary from about 28° F. (−2° C.) along the ice caps to above 86° F. (30° C.) in tropical seas. The lower limit is controlled by the formation of ice, which occurs at a lower temperature in sea water due to its salinity. The upper limit is determined by the amount of radia-

ISOTHERMS, the lines of equal temperature, would be parallel if the waters of the oceans did not circulate. But, because of ocean currents (p. 52), isotherms are irregular, as shown here, with only slight shifts northward or southward as the seasons change. Sea water's specific heat (amount of heat required to raise the temperature 1°C.) is so high that the daily temperature change of oceanic surface water is only about 0.2° C.

(see temperature key above)

OCEAN SURFACE TEMPERATURES IN FEBRUARY

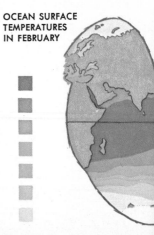

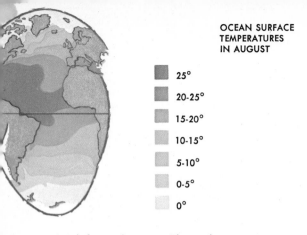

25°

20-25°

15-20°

10-15°

5-10°

0-5°

0°

tion received from the sun. Thus, the temperature
of the surface water is warmest in tropical seas.
In the Persian Gulf, the surface water temperature
may reach 96° F. (35.6° C.). Sea water cools gradually toward the poles, where the sun's rays strike the
surface at a greater angle.

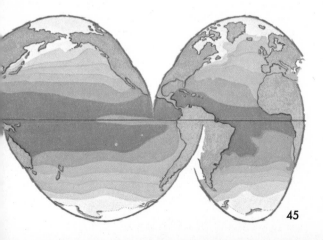

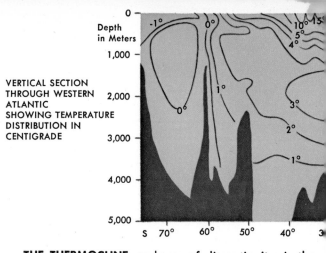

VERTICAL SECTION THROUGH WESTERN ATLANTIC SHOWING TEMPERATURE DISTRIBUTION IN CENTIGRADE

THE THERMOCLINE, or layer of discontinuity, is the region where the relatively thin layer of water that has been warmed by the sun's rays meets the deeper cold water. Very cold water is much more dense than warm water even with a higher salinity. Due to differences in density, these two layers of water do not mix easily, hence there is a sharp temperature change—the thermocline—located at the 50° F. (10° C.) isotherm. The thermocline is at the surface at 50° to 60° N. and S. latitudes, deepest in temperate regions (1,950-2,600 feet or 600-800 meters) and shallow in the tropics (300-1,600 feet or 100-500 meters). Nearly all plants and many animals of the sea live in the warm upper layer.

Below the thermocline, the temperature decreases rather slowly but regularly. In most oceans, the temperature on the bottom is about 35°-37° F. (2°-3° C.), but in great depths it may drop to nearly 28° F. (−2° C.).

46

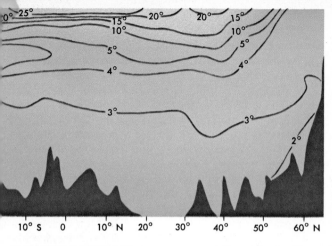

10° S   0   10° N   20°   30°   40°   50°   60° N

PRESSURE IN THE SEA increases rapidly with depth due to the density of the water. Air pressure at sea level is about 14.5 pounds per square inch, or one atmosphere (1.02 kgs/cm²). Even at an altitude of 15,000 feet (460 meters), the pressure changes very little. But in the sea, the pressure increases at one atmosphere for every 33 feet (10 meters) of depth.

DECIBARS are units used by oceanographers to measure pressure in the sea. A decibar (0.1 of a bar) is equal to the pressure of 1 meter of sea water on 1 square centimeter (about 1.4 pounds). A bar—ten times greater—is equal to approximately one atmosphere.

Pressure in the sea ranges from 0 decibars at the surface to more than 10,000 decibars (14,000 pounds, or 7 tons) at great depths. The great pressures in the depths have some

effect on sea-water density, but the change is slight, as water is almost incompressible.

Oceanographers' instruments must be capable of withstanding pressure changes from the surface to the bottom of the sea. One of the problems facing submarine designers is constructing hulls strong enough to resist the great pressures. The loss of the submarine *Thresher* in 1963 was probably due to implosion, the inward collapse of the submarine compartments.

47

DENSITY OF SEA WATER may vary so slightly that differences may cause a sinking or a rising of whole water masses. Such differences are one of the basic causes of ocean currents. Thus, the density of sea water plays a highly important role in the oceans.

Pure distilled water at 4° C. at atmospheric pressure is the base for determining the density of sea water. If the water is warmed, its density changes because of expansion; if the water is cooled, its density changes due to contraction. Adding salt increases the density, and compressing the water also increases its density slightly. These three factors—temperature, salinity, and pressure—determine water density.

The density of sea water at the surface (atmospheric pressure) can be determined simply by measuring the temperature and salinity. Below the surface, it is necessary to know also the pressure (depth in meters is equal to the pressure in decibars). With these facts, the density can be calculated from published tables. Such a table shows, for example, a density of 1.028 g/cm³ for a salinity of 35 ⁰/₀₀ (parts per thousand grams of water) and a temperature of 0° C. The density will be higher or lower than 1.028 as the salinity, temperature, and pressure are varied.

In making their calculations, oceanographers drop the full figure (1.0) and write the density as 28.00. And the density of sea water at atmospheric pressure is usually written as σt (sigma-t), as shown on p. 49. The density of sea water measured against the density of distilled water at 39.2° F. (4° C.) is its specific gravity; thus, the specific gravity and the density of sea water are numerically equal.

48

WATER MASSES in the oceans differ in characteristics, and because of their different densities they tend not to mix. In each water mass, the temperature and salinity vary from the surface downward. The temperature and salinity may be plotted in relation to depth, or, as shown here, they may be plotted against each other to show the characteristic curve of each water mass. Called T-S diagrams, these curves are a convenient way of indicating the differences in water masses.

If the σt-curves (p. 48) are plotted on the T-S diagrams, the stability of the stratification between waters of the upper and lower levels is indicated from the relationship of the slope of the T-S curve and the slope of the σt-curves.

A PROFILE OF THE GULF STREAM has been constructed below. At the left, the observed temperature and salinity have been plotted against depth. On the right, these have been plotted against each other to construct a T-S diagram. The density curves indicate, by the angle at which they cross the T-S line, the approximate stability of the water mass. Water masses are identical even at different depths.

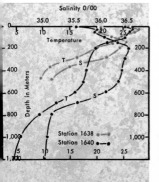

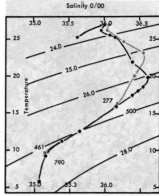

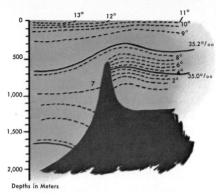

Depths in Meters

AN OCEANIC RIDGE divides the two open basins shown here, and the water masses on each side of the ridge are sufficiently different so that they support different kinds of fishes. In the English Channel, for example, two water masses may be distinguished by the different species of arrowworms present.

DIFFERENCES IN WATER MASSES are initiated at the surface, where the water is subjected to different rates of evaporation and to heating or cooling by the atmosphere. These surface waters retain their characteristics even after moving a long distance from the area of formation, until finally they become mixed with other water masses.

Water masses can be traced and identified by their unique features as they are moved by currents from their point of origin. They may travel thousands of miles through the open ocean or move into deeper or into shallower areas. Antarctic Intermediate water can still be distinguished in the Caribbean off South America because of its distinctive T-S curve. In the Florida Current, off Miami, three water masses can be distinguished by their T-S curves: Gulf of Mexico water, Yucatan Channel water, and oceanic water. Each has attained its nature from the area of its origin. By the time these have reached Cape Hatteras, they are combined by mixing to form the typical Gulf Stream water mass with an entirely different T-S curve.

Water masses vary also in chemical makeup and hence in productivity. Often two water masses that are nearly barren may be highly productive when brought together to form a new mass. The original masses may have each lacked particular trace elements necessary for planktonic life.

Biologists are interested in water masses because each separate mass may be inhabited by its own type of planktonic animals, called "indicator species." These animals are a means of identifying water masses without constructing T-S diagrams.

COLOR differences of the water at the entrance to the English Channel were detected from the air by mackerel spotters in 1923. Later these were determined to be inhabited by different species of arrowworms— *Sagitta elegans* in blue water and *S. setosa* in green water. The variation in population densities is shown in different seasons in 1934 and 1935.

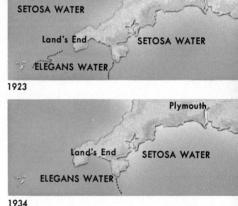

Plymouth

SETOSA WATER

Land's End

SETOSA WATER

ELEGANS WATER

1923

Plymouth

Land's End

SETOSA WATER

ELEGANS WATER

1934

Plymouth

Land's End

SETOSA WATER

ELEGANS WATER

1935

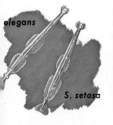

elegans

S. setosa

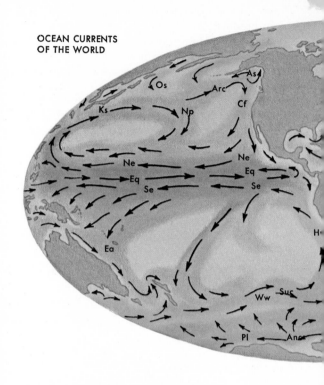

MAJOR CURRENTS in the ocean are of two general types: vertical and horizontal, both surface and deep. Lesser and variable currents may be caused by tides and storm conditions. Some currents are of short duration and cover only a small area; others, such as the great oceanic circulation systems, are permanent. The map above shows the world's major surface current systems.

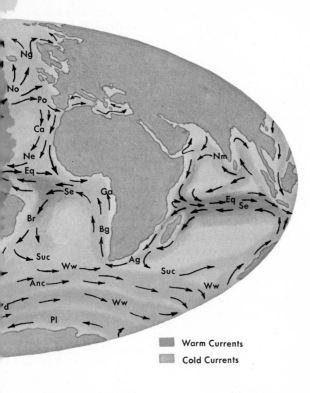

Warm Currents

Cold Currents

Ag—Agulhas	Eq—Equatorial	Nm—Northeast Monsoon
An—Antilles	Fa—Falkland	No—North Atlantic
As—Alaska	Fl—Florida	Np—North Pacific
Bg—Benguela	Ga—Guinea	Os—Oyashio
Br—Brazil	Gu—Gulf	Pl—Polar
Ca—Canaries	Hu—Humbolt	Po—Portugal
Cf—California	Ks—Kuroshio	Se—South Equatorial
Ch—Cape Horn	La—Labrador	Wd—Weddell
Ea—East Australia	Ne—North Equatorial	Ww—Westwind drift
Eg—East Greenland	Ng—Norwegian	

Current limits: Anc—Antarctic Convergence; And—Antarctic Divergence; Arc—
Arctic Convergence: Suc—Subtropical Convergence

MAJOR OCEAN CURRENTS are basically produced by two factors: distribution of density and effect of wind stress on the sea surface. Variations in density are widely distributed due to differential heating and evaporation. They cause the waters to move both horizontally and vertically. Major ocean currents are set in motion by these variations and by the drag of the wind on the surface layers. Once the water mass begins to move, it is deflected due to Coriolis force, and the major surface current circulation patterns are established. Upwelling and sinking of water masses are similarly caused by density differences and the blowing away or piling up of waters under the stress of air currents. Density gradients so slight that they are difficult to measure may nevertheless be sufficient to produce or to maintain ocean currents.

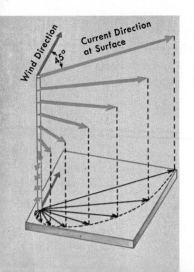

CORIOLIS EFFECT, due to the earth's rotation, causes objects on the earth's surface to be deflected to the right in the Northern Hemisphere and to the left in the Southern Hemisphere. Thus water is moved onshore by winds coming in at an angle. Surface currents in the Northern Hemisphere are deflected 45° to the right of the wind direction; in the Southern Hemisphere, 45° to the left. At the depth at which the current speed is 1/23 of the surface speed, the current moves exactly opposite to the direction at the surface, a phenomenon called the Ekman spiral.

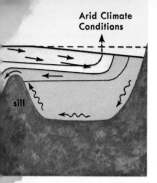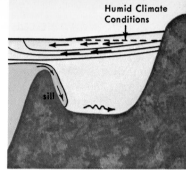

In the Mediterranean Sea and the Red Sea, an arid atmosphere causes great evaporation. Surface water flows in to replace the loss, and the heavy, dense water flows out over the high sill. On the right, the system is reversed due to heavy rainfall.

Surface circulation of the oceans (pp. 52-53) is broadly similar in both the Atlantic and the Pacific and is in opposite directions on either side of the equator. In the North Atlantic, the wind, circulation at the area of the Northeast Trade Winds moves the surface waters to the west due to the Coriolis effect. The water moves westward as the broad North Equatorial Current and is deflected to the right, entering the Caribbean Sea and the Gulf of Mexico. In the Gulf, a slight hydrostatic head (about 1 foot or .3 meters) is formed, and the water rushes through the Straits of Florida as the Florida Current, traveling at a speed of about 4 miles (7 kilometers) an hour. Outside the Straits, it joins the Antilles Current to form the Gulf Stream. Part of the Gulf Stream is deflected into the Arctic Ocean as the Norway Current. The remainder moves southward along the European coast as a slow-moving mass, returning to its point of origin in the tropics.

55

A DEEP-WATER CIRCULATION from north to south also occurs in the Atlantic. To the west of the Gulf Stream below Greenland is an area of considerable turbulence where the warm, highly saline waters of the Gulf Stream are cooled, the density increases, and the water sinks, starting the process of convective turnover. The cold water mass sinks to intermediate levels where it is displaced to the south. In the tropical regions of the North Equatorial Current, the surface waters are in turn heated, evaporated, and displaced by wind stress. Here the deep waters rise to the surface, completing the cycle.

Movement of the deeper water masses was unknown until deep-current meters, buoys and radioactive trace-element detectors came into use. One of the great events of sea-current research was the discovery and plotting, in 1952, of the Cromwell

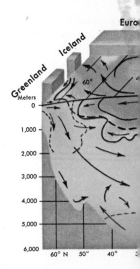

SURFACE and deep-water circulation in the Atlantic Ocean are shown here schematically. Warm waters are shown in red, cool waters in light blue, and cold waters in dark blue. The direction of the currents is indicated by arrows. Solid lines on the surface represent areas where the currents of different temperatures converge. Solid lines beneath the surface connect waters of equal salinity.

Current in the equatorial Pacific by scientists from the Scripps Institution of Oceanography. This great current, about 250 miles (400 kilometers) wide and coming to within a few meters of the surface in some places, sweeps for thousands of miles in an eastward direction at a speed of 3.5 miles (6.5 kilometers) per hour. A similar current has been found between Africa and South America—the Lomonosov Current.

Tidal currents are horizontal. As the level of the ocean becomes lower in relation to a bay or lagoon, the water in the bay flows outward as a tidal current. Coastal tidal currents that parallel the shore change directions with the tides. Strong, persistent winds may force normal tidal currents into reverse. When wind currents and waves meet tidal currents head-on, dangerous seas may be generated.

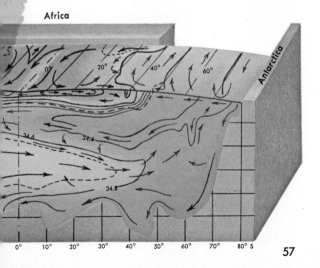

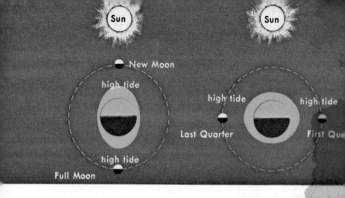

TIDES are the alternate rise and fall of water in the oceans and their tributaries under the influence of the moon and, to a lesser degree, the sun.

Water on the side of the earth facing the moon is drawn by lunar attraction slightly toward the moon. At the same time the earth is pulled slightly away from the water envelope on the opposite side. Between these two masses the water is drawn toward the earth by its gravitational field. The result is two bulges (high tides), one toward the moon and the other away from it, with a flattening on either side (low tides).

The effect of the moon is greatest but the sun also moves water particles. When the sun and the moon are in a straight line with the earth (new moon and full moon), the tides are at their highest and lowest. These are the *spring tides*. When the sun and the moon are at right angles with relation to the earth (first and third quarters), the water movement is least. These are the *neap tides*.

Two high tides and two low tides (semidiurnal) are produced every 24 hours, as the earth rotates around its axis. But since moonrise occurs about 50 minutes later each day, each high tide actually occurs about 25 minutes later than the one before, and a corresponding tide on the next day will be 50 minutes later.

DIURNAL TIDES, a high and a low occurring only once every 24 hours, are found in parts of the China Sea and in parts of the Gulf of Mexico. · Mixed tides, in which the alternate highs and lows are markedly different, occur in many places, too, as in parts of Australia and in certain areas along the U. S. Pacific coast.

The height of the tide varies with geography but is greatest at the shore end of bays, in gulfs, and in areas where tides must move through a constriction. The greatest range of the tides—about 54 feet (16 meters)—occurs in Minas Basin at the head of the Bay of Fundy in Nova Scotia. Due to the peculiar configuration of the land, the tidal currents run inward only 4 hours and outward 8 hours, rather than the usual 6 hours each way. In open oceanic areas, such as at Tahiti or at Key West, Florida, the tidal range may be 2 feet (.6 meters) or less.

In the open ocean, the time of the tide may occur almost exactly at the time the moon passes overhead. At other places, because of the basin topography, there is a considerable delay. These variable features cause the great difference in times of high tide along any coast.

Tidal currents are caused by the water rushing through constricted areas to fill bays, sounds, gulfs, or rivers. They alternate in direction as the tides change. .

Nova Scotia's Bay of Fundy at high tide (left) and low tide (right).

WAVES occur everywhere in the oceans, except in those rare times when the sea's surface is motionless during long periods of calm. But even on calm days it is not uncommon for large waves to pound the shore, seeming to roll up out of the ocean from nowhere, caused by storms many miles away.

Surface waves are caused by winds. Over even small areas, the winds are unequal in force, and so first ripples and then wavelets build up on the sea's surface. If the wind continues to blow with sufficient force and there is enough distance, waves are formed. The wind exerts two forces: a direct force on the back side of the wave, and an eddy or suction on the front side.

Theoretically, a wave should be a smooth curve in cross section. A series of these waves should be long and parallel to each other. Such waves appear

Diagram at left below shows parts of a typical wave. In illustration at right, note the inward bend of seaweeds on each side of wave, indicating the orbital movement of the water.

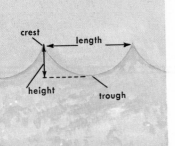

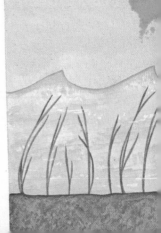

only near the shore and occasionally in hard storms. In the open sea the surface is very irregular, with short waves moving in the midst of many others. In moderate seas large, fairly regular swells may be interspersed with numerous steep seas.

A wave's height, period (length of time for two successive crests to pass a given point), and length are determined by the speed and duration of the wind and by the distance (fetch) the wave has traveled. Large waves travel several hundred miles. During great storms, oceanic waves may reach a height of 49 feet (15 meters), a length of just over 1,000 feet (310 meters), and a period of about 14 seconds or a speed of 55 miles (89 kilometers) per hour. To cause such large seas a fetch of about 500 miles (900 kilometers) and winds in excess of 70 miles (130 kilometers) per hour are necessary.

Note that the skiff remains in one position while the wave travels beneath it. The wave moves, but the water does not, as is shown by the skiff's stationary position.

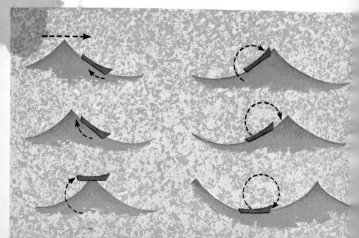

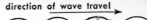

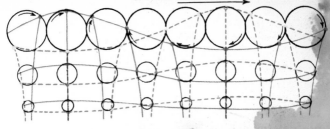

Solid red lines show the form of the wave and the position of
particle at particular instant. Dotted green lines show
shape of wave and position of particle half an orbit later.

AS A WAVE MOVES FORWARD, the surface water
itself remains almost stationary while the particles
of water within the wave travel in a circular motion. In a wind wave, the up-and-down motion at
the surface may be very great, but the motion decreases with depth until finally the wave effect is
lost. In hard storms, however, the motion may be
great enough to stir the bottom and tear loose
sponges and coral from several hundred feet deep
and wash them up onto the beaches.

As a wave nears the beach and the water becomes shallower, the lower part of the wave begins
to drag on the bottom. This causes the wave to rise
more steeply, as the top moves faster than the bottom and finally falls forward as a breaker. Great
storm waves breaking on a headland may send
sprays of water a hundred feet into the air. Break-
ers striking a shore may equal the pounding by
many hundreds of tons of water and do great damage. Waves approaching a coast at an angle are
turned as the first part of the wave slows down
when the water shoals. This causes the waves to
approach the shore almost parallel to it.

Seiches are long waves that occur in lakes, bays, sounds, and even on some open coasts. Actually oscillations, they are caused by changes in barometric pressure and other factors. Once started, they may continue for a long time. Internal waves, which are of particular interest to oceanographers, are formed beneath the surface at the interface of two layers of water of different densities. In polar seas, where the surface water is lighter, a slow-moving ship may set up internal waves. Though they are not seen on the surface, they may actually constitute a drag on the hull of the ship. If the ship's speed is increased, the internal waves are not formed and the drag stops. Deep internal waves have been found in many parts of the ocean.

Refraction of waves breaking around a headland

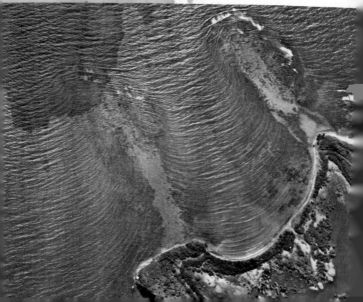

DESTRUCTIVE WAVES may originate from earthquakes that set up longitudinal oscillatory waves, which travel at the speed of sound and break the surface with reports like thunder or gunfire. A ship struck by one of these shock waves may be rocked violently. Sometimes these waves are accompanied by the release of great volumes of gases that, as they break the surface, send out such large transverse waves that they damage ships. Fortunately, waves of this kind are rare.

Tsunamis, also called tidal waves, are the most destructive waves known. The waves are started by submarine landslides during earthquakes or by the eruption of submarine volcanoes. The distance between the crests of these large transverse waves may be 500 miles (800 kilometers), with periods ranging from 10 to 60 minutes. In the open ocean, a tsunami's speed is determined by the depth of the water. In 500 feet (150 meters) a wave may travel at 75 knots (140 kilometers per hour), while in 15,000 feet (4,600 meters) it may travel at 420 knots (780 kilometers per hour).

Tsunamis usually cause great damage at or near their point of origin. They may spread out and be unnoticeable in the open sea but become destructive again on the approach to land thousands of miles away. The great earthquake in Lisbon on November 1, 1755 caused a destructive tidal wave in that city. The wave was still 13 to 20 feet (4 to 6 meters) high when it reached the West Indies. Japan has suffered from many such tsunamis.

One of the greatest tidal waves known was caused by the explosion in 1883 of Krakatoa, a small volcanic island off Java and Sumatra. More than 36,000 persons were killed as the waves in some regions attained a height of about 115 feet (35 meters). The waves were recorded 32½ hours later in the English Channel, having crossed the Indian Ocean and traversed the length of the Atlantic. In 1960, an earthquake on the Chilean coast caused a severe tsunami in Hawaii and Japan.

Unrelated to tsunamis are storm-caused waves, or inundations, resulting from the pile-up of water along a shore under the force of extreme winds. A very low barometric pressure at the same time may cause the sea to rise as much as 16 to 20 feet (5 to 6 meters) above mean high water. Galveston, Texas, experienced a disaster of this type in 1900 when the water during a hurricane rose about 16½ feet (5 meters) and drowned nearly 6,000 people. The Mississippi hurricane of 1969 had a wave crest of approximately 30 feet (10 meters). Occasionally, due to storms at sea, a "lens" of high water may occur on a coast unaccompanied by high winds.

Reproduction of "The Great Wave" by Hokusai

SEA ICE is formed by the freezing of sea water at the surface. According to the temperature at which the ice was formed, the rapidity of formation, the age of the ice and other factors, sea ice may be as salty as sea water, with the salts contained either in cells of brine or as salt crystals.

Sea ice is found in many forms, but essentially it is flat pack ice that may be sculptured by physical forces into great flat floes, broken floes, irregular shaped masses, or pancake ice. Sea ice in the Arctic Sea in the winter is about 10 to 13 feet (3 to 4 meters) thick; in summer, it is 6½ to 10 feet (2 to 3 meters) thick. In the Southern Ocean, Antarctic Sea ice usually consists of vast flat floes, but in the Arctic Ocean, where the movement of the ice is restricted by land masses, the ice is thrown into pressure ridges and hummocks underneath which it may be very thick. The ragged surface of the sea ice in the Arctic is a characteristic feature.

ICEBERGS originate from glaciers on land and consist of fresh-water ice. In the Antarctic, icebergs are formed along the edge of the shelf ice extending into shallow water from the continent. Great flat-topped masses break off and float seaward through the pack ice. Often, because of oceanic currents, they move in the opposite direction from the pack ice itself. These masses of ice, so large that they were once thought to be islands, may be 12 to 18 miles (20 to 30 kilometers) wide, 62 miles (100 kilometers) long, and as much as 2,620 feet (800 meters) thick. Some are so large that whaling fleets may take refuge in their lee during the severe storms that occur in the Antarctic.

Arctic icebergs, jagged and seldom with flat tops, are formed mainly from the Greenland glaciers and along the shores of the Bering Strait. The East Greenland and Labrador currents sweep them as far south as the Grand Banks where they represent a hazard to shipping. After the sinking of the *Titanic* in 1911, the major nations of the North Atlantic banded together to form the International Ice Patrol. One of the duties of the United States Coast Guard is to locate icebergs and warn shipping of their position and to predict their occurrence and drift by means of hydrographic data.

A TYPICAL ICEBERG in Antarctic is tabular or flat-topped (right), broken from a shelf of ice. Most icebergs in Arctic are jagged spikes (below).

The illustration shows that more than 80 percent of the iceberg is concealed below the surface. A ship can strike the submerged portion of an iceberg while it is still a reasonable distance from the portion protruding above the water.

CHEMICAL OCEANOGRAPHY

The chemical composition of the open sea water varies only slightly from ocean to ocean, but near land the composition may vary greatly due to runoff or river waters. It varies also with depth.

Oceanographers attempt to determine the amounts of various chemicals in different water masses and the degree and nature of seasonal variations. They want to find out the sources of the chemicals, how they are distributed and assimilated, and what effect marine organisms have on their amount and distribution—horizontally, vertically, and seasonally.

Chemical oceanographers filter a sea water sample (left) to determine the levels of particulate radionuclides present. At right, they are adjusting the flow valves of a PCO_2 analyzer.

NEARLY ALL OF THE ELEMENTS known to man are found in the oceans. Minerals are being added constantly from land sources, but the composition of the original seas was probably much like today's.

Of the many chemicals dissolved in the seas, the most obvious, at least to taste, are the salts of sodium, magnesium, calcium, potassium, and strontium, occurring as chlorides, sulfates, or bromides. These major elements in the sea occur in a constant ratio of one to another. The minor elements fluctuate in amount both geographically and seasonally, due mainly to variations in biological activity, ice formation, and river run off. Even small traces of these minor elements may be important to plants and animals, however.

Each cubic meter of sea water contains about 0.004 mg. of gold. This means there are about 2,000 atoms per cubic millimeter, enough so that many atoms may contact the surface of a diatom or a piece of detritus. This helps explain why some animals are able to concentrate rare elements in significant amounts. Tunicates, for example, contain large amounts of the rare vanadium.

ELEMENTS IN SEA WATER (symbols for elements shown on chart below)

Aluminum (Al)	Cerium (Ce)	Gold (Au)	Nickel (Ni)	Silver (Ag)
Antimony (Sb)	Cesium (Cs)	Iodine (I)	Nitrogen (N)	Sodium (Na)
Arsenic (As)	Chlorine (Cl)	Iron (Fe)	Oxygen (O)	Strontium (Sr)
Barium (Ba)	Chromium (Cr)	Lanthanum (La)	Phosphorus (P)	Sulfur (S)
Bismuth (Bi)	Cobalt (Co)	Lead (Pb)	Potassium (K)	Thorium (Th)
Boron (B)	Copper (Cu)	Lithium (Li)	Radium (Ra)	Tin (Sn)
Bromine (Br)	Deuterium (D)	Magnesium (Mg)	Rubidium (Rb)	Titanium (Ti)
Cadmium (Cd)	Fluorine (Fl)	Manganese (Mn)	Scandium (Sc)	Vanadium (Va)
Calcium (Ca)	Gallium (Ga)	Mercury (Hg)	Selenium (Se)	Yttrium (Y)
Carbon (C)	Germanium (Ge)	Molybdenum (Mo)	Silicon (Si)	Zinc (Zn)

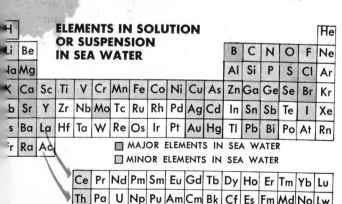

ELEMENTS IN SOLUTION OR SUSPENSION IN SEA WATER

☐ MAJOR ELEMENTS IN SEA WATER
☐ MINOR ELEMENTS IN SEA WATER

SALTS from the sea have been used by man for thousands of years. Primitive man scooped crystals of salt from rock pockets along shores where the sea water had evaporated. Later, he built shallow ponds by the sea, flooded them, and then let the water evaporate to leave the salt deposits.

Scientists used 77 sea-water samples collected during the cruise of H. M. S. *Challenger* (1872-76) from all oceans and depths to make a comparison of their salt content. Analyses showed a nearly constant ratio of major elements in sea water.

The level of salt content at which sea water becomes brackish and brackish water becomes fresh is not agreed upon. Some of the fresh-water lakes and springs of Florida have a rather high salt content, for example. Partly protected arms of the sea usually have a noticeably different salinity from that of the open ocean. Fresh-water runoff from rivers and streams or underground seepage of ground water may lower the salinity. If there is little or no runoff, the salinity is generally high; in other words, arid conditions exist, with a high rate of evaporation. The water goes into the atmosphere, and all

Percentage Composition of Salts in Ocean Water	
Sodium	30.4
Magnesium	3.7
Calcium	1.16
Potassium	1.1
Strontium	0.04
Chloride	55.2
Sulfate	7.7
Bromide	0.19
Boric acid as H_3BO_3	0.07

OCEANOGRAPHERS give two names to the saltiness of the sea; salinity and chlorinity. Salinity refers to the total amount of salts—approximate dry weight of salt in 1,000 grams of sea water. In areas unaffected by melting ice in polar regions, the salinity of the sea varies little from about 35 parts of salt to 1,000 grams of water (indicated by the

Salt is obtained by evaporation in these ponds in Bahama Islands.

of the salts remain in the sea. The Red Sea, for example, is surrounded by the arid deserts of the Arabian peninsula and eastern Africa. As the dry winds heated by the tropical sun pass over the sea, large amounts of water are evaporated. As a result, the salinity is about 40 $^0/_{00}$.

As its salinity increases, water becomes heavier and tends to sink. The colder it is, the more rapidly it sinks. When highly saline water from the tropics is carried by currents, such as the Gulf Stream, into cold areas and is cooled, it sinks slowly.

symbol $^0/_{00}$). Salinity is difficult to determine directly. A gravimetric reading, a measure of density, can be taken with a float, or a volume-weight measurement can be made. But usually the salinity is calculated from chlorinity determination.

Chlorinity, the halide content of 1,000 grams of water, can be determined very ac-curately. Chlorinity is proportional to the salt content, so the resultant salinity can be determined by multiplying chlorinity by 1.80655. The result is expressed as $^0/_{00}$ (parts per thousand). If very accurate procedures are maintained, salinity as slight as about 3 parts in 35,000, or 35.003, can be determined. Salinity determinations are accurate to $0.003^0/_{00}$.

ELEMENTS occur in varying amounts in the oceans. Some animals concentrate rare elements in their tissues or fluids. Others absorb the elements in their outer membranes, commonly using them to make skeletons. The amounts of phosphates and nitrites-nitrates fluctuate widely, based on their utilization by plants and by their release due to the action of microorganisms, mainly bacteria, during decomposition. These cycles (nitrogen, phosphorus, and others) are of great importance to life in the sea.

NITROGEN passes through a cycle in the oceans. Organic nitrogen in plant cells or in the stomachs of animals that have eaten the plants is partially broken down into ammonium (NH_4) either by bacteria or by protein-destroying enzymes in animal digestive fluids. The ammonium can be utilized directly by plants or may be oxidized slowly into nitrites (NO_2) and then changed into nitrates (NO_3) by bacteria. These changes occur just below the photosynthetic zone, and in shallow seas, either directly above the bottom or in the bottom deposits. Most plants can absorb nitrogen in all three forms—ammonium, nitrites, and nitrates. But most soluble and particulate nitrogen compounds in animal wastes are first broken down by bacteria.

Nitrogen gets into the oceans also from river runoff and from rain. Rain may contribute annually about 28 mg. of nitrate nitrogen and 56-240 mg. of ammonium nitrogen to each square meter of sea surface.

NITROGEN CYCLE ↓↓↓

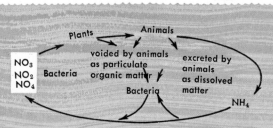

Plants → Animals

NO₃ NO₂ NO₄   Bacteria

voided by animals as particulate organic matter

excreted by animals as dissolved matter

Bacteria

NH_4

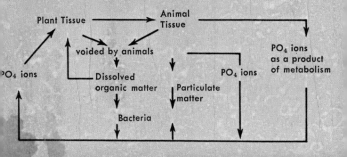

PHOSPHORUS is one of the most important minerals in the sea for the growth of plants but is not always present in sufficient quantities. It occurs in two forms: as phosphates (PO_4) and as dissolved organic phosphorus. Both forms are absorbed by plants and utilized in growth or stored.

Most of the organic phosphorus in plants is released as phosphates when the plants die. Phosphorus in phytoplankton eaten by zooplankton herbivores is converted into phosphates and excreted. Parts not digested pass out of the gut in fecal pellets. Some of the phosphorus in these pellets is dissolved in the sea water as organic phosphorus. Some is converted into phosphates while still in the gut.

At each stage, the phosphorus moves to a deeper level in the sea, eventually becoming unavailable. Turbulence caused by storms will bring phosphorus back into the surface layers in shallow waters. It rises from deep water only in areas of upwelling. Without this replacement, the seas would soon be barren of life.

In tropical seas, where plant growth is about the same throughout the year, the amount of phosphorus remains nearly constant. In temperate regions, the amount of phosphorus varies, generally being less prevalent during phytoplankton blooms and more plentiful between blooms as it accumulates due to death and decay of the plants.

The cycle illustrated above shows how the phosphorus is distributed vertically through biological systems. The North Atlantic is not as rich in phosphorus as is the southern ocean. This vast reservoir of nutrients, through the vertical and horizontal movement of water masses, enriches the North Atlantic waters.

73

CARBON DIOXIDE AND CALCIUM play basic roles in the growth of both plants and animals. They also form large deposits, either inorganic or organic in origin, on the ocean floor. Together they are the major ingredients of the hard parts in the skeletons of starfishes and their kin and also of mollusk shells. Two forms of calcium carbonate occur: calcite, present particularly in cold waters; and aragonite, typical of warm waters.

Calcium precipitates from sea water most readily when the temperature and salinity are high and the carbon dioxide concentration is low. These conditions are found over shallow banks of tropical seas. Here the sun heats the entire water layer; evaporation is high, and the carbon dioxide level is low due to rapid plant growth. As a result, the pH (see below) rises and the water, now supersaturated with calcium, precipitates calcium carbonate, restoring the

pH. Bacteria, calcareous green algae, coralline red algae, corals, mollusks, and a host of other marine animals and plants precipitate calcium from water organically.

Carbon dioxide is derived from gas transfer at the surface and from metabolic activity. Carbon dioxide is chemically transformed to bicarbonate and is utilized as such by plants in photosynthesis. Complex carbon compounds are converted to carbon dioxide through bacterial decomposition and oxidation by dissolved oxygen. Carbon dioxide is probably not a limiting factor in the seas, and its abundance varies with the amount of biological activity.

Associated with the amount of carbon dioxide is the pH, or hydrogen ion concentration, of the water, an indication of alkalinity or acidity. The pH of sea water does not ordinarily vary greatly, hence has little effect on ocean life. Surface waters of the open ocean are alkaline, with a pH of 8.1 to 8.3. The pH drops in deeper water, and in the ocean middepths, where there are no photosynthetic plants, carbon dioxide concentrations increase, and the pH shifts below 7.0. Here, calcium carbonate dissolves. For this reason, limy shells are not commonly found in deep bottom oozes. Instead there are mainly silicate shells of radiolarians (p. 107) and diatoms.

Coral reef of calcium carbonate

This sulfur mine off Grand Isle, Louisiana, has three drilling-producing platforms, a heliport, living quarters, and a power plant. It can withstand hurricane winds and seas.

SULFUR, required in the life processes of all plants and animals, occurs in ocean water as sulfates, sulfites, and sulfides. Plants utilize sulfur in the form of sulfates, which are produced by chemical or by biological oxidation. When plants and animals decompose anaerobically, hydrogen sulfide is released. The odor of "rotten eggs" along mud banks and in marshes indicates a high concentration of this gas. It may be produced chemically or by the action of heterotrophic bacteria (pp. 102-103) in reducing proteins. Even inorganic sulfate compounds are reduced to hydrogen sulfide by these bacteria.

High concentrations of hydrogen sulfide in deep basins where currents are lacking are accounted for by the ability of heterotrophic bacteria to bring about oxidation in the absence of dissolved oxygen.

Hydrogen sulfide is hostile to all life other than bacteria. For instance, in the Black Sea there is little macroscopic life below about 590 feet (180 meters), the depth at which hydrogen sulfide is abundant.

Not all bacteria in the sulfur cycle produce hydrogen sulfide. The true sulfur bacteria produce pure sulfur and pass it back into the sulfur cycle by the process of oxidation ($2H_2S \mp O_2 2H_2O + 2S$).

75

OXYGEN in sea water is chemically bound with the hydrogen (H₂O). All of the free oxygen dissolved in the oceans either is absorbed from the atmosphere at the surface or is a by-product of photosynthesis in the shallow depths where light penetrates. During photosynthesis, water is split apart, releasing oxygen. Therefore, surface layers are richest in dissolved oxygen; the deep layers are poorest (except where oxygen-rich waters sink into the depths, as in Antarctica). Oxygen is lost into the atmosphere at the surface and is also consumed in respiration by plants and animals, in chemical oxidation, and in decomposition of organic material by bacteria.

AN OXYGEN MINIMUM layer occurs at a depth of from about 1,300 to 3,300 feet (400 to 1,000 meters) over a wide area of the ocean from about 20° N. to 20° S. latitude. In the tropics, the oxygen concentration may drop as low as 0.5 ml/l. This low-oxygen layer, deprived of oxygen by an excess of respiration by animals and by decomposition, is believed to originate at the surface in high latitudes; it then slips under the surface and moves toward the equator.

A tongue of water, about 1,900 to 2,700 feet (600 to 800 meters) thick and as much as 3,000 miles (4,800 kilometers) long projecting into the Pacific off Central America, has an oxygen concentration of only about 0.25 ml/l, too low for all except specially adapted marine life. Another such area is the Cariaco Trench in the lower Caribbean. The anaerobic waters support only bacteria. The Black Sea's deep waters lack oxygen but contain concentrations of hydrogen sulfide.

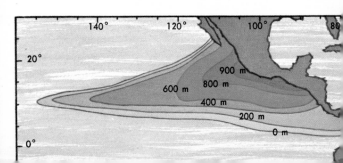

CONCENTRATIONS of oxygen in the sea may vary from about 9 ml/l (milliliters per liter) to nearly 0. (At sea level the atmosphere contains about 200 ml/l of oxygen uniformly distributed.) Cold water holds more oxygen than warm; thus, in tropical waters the amount of oxygen at saturation is about 4.5 ml/l, while in polar regions it is about 8.0 ml/l. Supersaturation may occur when photosynthesis by phytoplankton is very high. Polar waters are often supersaturated by 10 percent or more. During heavy plankton blooms in San Francisco Bay, supersaturation may reach 150 percent; in the shallow waters of coral reef flats, supersaturations of more than 250 percent may occur. Phytoplankton use large amounts of oxygen at night in respiration, and by daylight the oxygen level may have dropped to a supersaturation of 18 percent. In exceptionally heavy blooms, all the nutrients may be used up; then the organisms die and decompose.

PARTICULATE ORGANIC MATTER in the sea—composed of dead and decomposing plants and animals —is organic matter that is large enough to be filtered out of sea-water samples. It reflects the amount of plant and animal life present and hence the productivity of the sampled area. Of the various methods of analyzing the material, the most commonly used is the carbon content of the ash. This gives a significant indication of productivity, as carbon is the most important element in the formation of protoplasm. Chlorophyll a determination gives an indication of photosynthesis levels, but particulate organic matter has been found to contain chlorophyll derivatives that complicate the use of this indicator.

Especially in early larval stages, some animals can utilize particulate organic matter. A gelatinous organic material, leptopel, occurs in large quantities in sea water. It has been found in the guts of fishes in early stages of development and is probably an important food.

DISSOLVED ORGANIC MATTER in the seas may constitute more than the total of all the phytoplankton. In 1907, Pütter theorized that marine invertebrates could obtain nourishment directly from these dissolved organics, but how this was done was not understood. Now, with new information available, similar theories are being brought forth. Animals cannot assimilate basic elements, such as carbon, but it is possible that some invertebrates and larval fishes can utilize directly the amino acids dissolved in sea water.

Determining the amount of carbon compounds present in sea water is difficult. The inorganic salts make analysis more complicated. Dissolved organics, in fact, is a broad term used for compounds of organic nature that will simply pass through a filter, as opposed to particulate matter and nannoplankton that can be filtered out easily. More investigations are needed in this area.

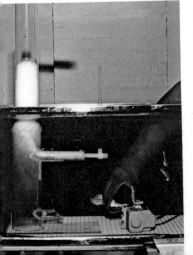

THE FOAM DOME (left) or protein skimmer is a device that pumps water in an aquarium through a pipe equipped with a venturi in which air is admitted. The air enters the water as minute bubbles that molecularly combine with the protein of dissolved organics from decomposing food or dead animals. These protein-laden bubbles rise in the dome as a viscous foam and are drained from the tank. Use of foam dome permits more animals to be kept together in an aquarium and eliminates necessity of frequent cleaning. Animals formerly thought too delicate to be held in tanks then survive.

METEOROLOGICAL OCEANOGRAPHY

Oceans play a significant role in determining the world's weather and climate. Thus, meteorologists are particularly interested in what happens where the sea and air meet and exchange heat and water vapor. Major air circulation patterns and their effect on ocean circulation are of great significance. The Doldrums (p. 86) and the "roaring forties" (p. 87), once so important to sailing ships, result from global air circulation. Hurricanes and typhoons are strictly marine, originating in the open ocean and dissipating over land where their great heat machines can no longer operate (p. 88). Because more than 70 percent of the earth's surface is covered by seas, land areas in the main only modify the world's climate or cause local irregularities.

HEAT INPUT IN THE OCEANS is one of the most important factors in the earth's weather and climate. Because of sea water's high specific heat (capacity to absorb and retain), the oceans are the earth's great heat reservoirs. In contrast, the specific heat of land is very low and heat absorbed during the day is quickly lost, often overnight. Thus, the oceans are much more effective than land as a thermostatic control of climate. For the earth as a whole, the total amount of heat received annually from the sun almost exactly equals the amount of heat lost by radiation and reflection.

The world's tropical regions lie much more directly in the path of the sun's rays than do the high latitudes. They absorb more heat from the sun than they lose by radiation and reflection. But in high latitudes, the reverse is true. Air currents and ocean currents move heat from low latitudes to high latitudes. In parts of the oceans, particularly in the North Atlantic, ocean currents are of considerable importance in this heat transfer.

Most of sun's rays striking the earth directly are absorbed; at an angle they are reflected.

Similarly, oceans control the moisture content of the atmosphere through the evaporation of water from their surface. In some areas of the tropics, vast cloud barriers stretch across the open ocean from one horizon to the other, marking the boundary of high evaporation. In the tropics, the warm layer in the oceans may be less than 300 feet (100 meters). It becomes deeper in warm temperate regions; it becomes thinner and disappears in high latitudes. Though much more heat is needed to raise water temperature than to raise land temperature, it is also true that the sea can release a great deal more heat to the air without greatly changing its own temperature. There is a lag of about two months between seasonal temperatures of sea and air.

81

WIND CURRENTS in the atmosphere are in some ways much like the currents in the sea. Shown at right is the wind system of a rotating hypothetical globe covered with water and surrounded by air.

The wind belts start with the Equatorial Low, a low-pressure band that circles the globe at the equator. Here the wet, moisture-laden air rises, causing rain and generally cloudy conditions. Toward the poles on each side are the Northeast and Southeast trade winds, where the winds blow steadily the year around with little variation. At about 30° N and and 30° S are the Subtropical Highs, or Subtropical Divergence. In this belt of descending air and high pressure, the winds are variable, but there are calms with clear atmosphere. From these highs toward the poles, the winds blow to the northeast and southeast—the so-called Prevailing Westerlies that rush into the Subpolar Low of 60° latitude. The rising air masses bring heavy precipitation. Toward the poles lie the Easterlies that move toward the Subpolar Low.

The Polar Highs are high pressure areas of sinking air. Exchanges of air occur also in the upper atmosphere, from the Equatorial Low directly to the Polar High. As this wind blows back toward the equator, it meets the Prevailing Westerlies at the Subpolar Low and forms a "polar front." Here the warm air from the south pushes up over the cold polar air, cools rapidly, and forms an area of unsettled weather. Much of the changing weather in the middle latitudes is derived from these fronts. Due to the earth's inclination and uneven distribution of land and water, the whole equatorial system is shifted slightly northward.

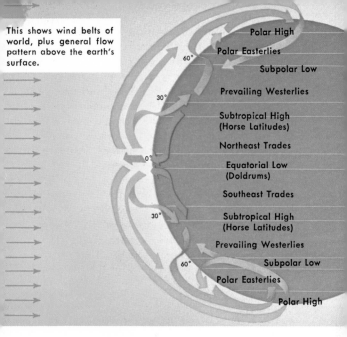

This shows wind belts of world, plus general flow pattern above the earth's surface.

Polar High
Polar Easterlies
Subpolar Low
Prevailing Westerlies
Subtropical High (Horse Latitudes)
Northeast Trades
Equatorial Low (Doldrums)
Southeast Trades
Subtropical High (Horse Latitudes)
Prevailing Westerlies
Subpolar Low
Polar Easterlies
Polar High

CONTINENTAL MASSES and other factors complicate wind systems. While the main system is a major factor in initiating the ocean currents and setting the pattern for the climate of the world, the deflection of the wind systems near the margins of continents and the resulting vagaries cause many of the erratic coastal ocean currents. The struggle for dominance between the air mass movements over the continents and those over the oceans are the cause of sudden changes in weather in such coastal regions as California, Maine, and Florida. The inclination of the earth's axis and the uneven distribution of land and water are factors also.

83

THE CLIMATES of the world's land areas, based on the air circulation system, should lie in parallel belts that coincide with the wind patterns, but this is not the case. Barometric pressure patterns are rather regular only over the sea; over land areas the air masses are heated and cooled unevenly, resulting in variable pressures and irregularities in the air-current patterns. Also, as sea winds come in over the land, they may lose all of their moisture due to differences in land temperature or to the interference of mountain ranges. Moisture-laden Southeast Trade Winds and winds from the Doldrums release heavy rains as they pass over Brazil. By the time they reach the eastern slopes of the Andes, they are wrung dry, and the western side of the mountains are barren, desert wastes. Thus, travel over land greatly alters the nature of the air masses.

Moisture-laden winds from the sea rise when they come to a land mass. The moisture is lost as precipitation, and the winds, now cool and dry, move down the other side of the slope.

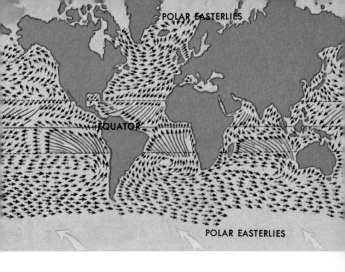

POLAR EASTERLIES

EQUATOR

POLAR EASTERLIES

EUROPE'S CLIMATE is affected to a great degree by the interaction of air and ocean currents. Trade winds drive the surface water of the oceans westward in two broad bands, the North and the South Equatorial Currents (p. 52). In the North Atlantic, the North Equatorial Current turns northward along with part of the South Equatorial Current and then, in the region of the Westerlies, turns back toward the eastern Atlantic as the West Wind Drift. Continental air masses contributing to the Westerlies in the North Atlantic are dry and cold. They bring about the "deep freeze" conditions of northeastern Canada, Newfoundland, and Labrador. When these same winds blow across the warm waters of the Gulf Stream, they pick up moisture and are heated by radiation from the sea surface. By the time they reach Europe, they are pleasant and mild.

In the Doldrums, a sailing vessel could only drift in the sea.

THE DOLDRUMS, also known as the Equatorial Low, comprise a belt located in the equatorial regions of the Atlantic and Pacific oceans in which there is no wind or only light and variable breezes that come first from one direction and then another. In February and March, the Doldrums are a small region concentrated on the west coasts of Africa and America just north of the equator. By July and August, the belt has moved northward, with its axis at about 7° north latitude. Now it is several hundred miles wide and extends nearly across both oceans. For weeks the surface of the sea may be completely calm or merely ruffled from one direction or the other. Heavy overcasts are frequent, with thunderstorms and cloudy, rainy skies a common occurrence, as the moisture-ladened Northeast and Southeast trade winds meet. Warmed under the direct rays of the sun, the trade winds expand and rise over a wide area. The result is low barometric pressures—hence the name Equatorial Low. Sailing vessels crossed in the narrowest section or waited until winter while the Doldrums were not well established.

THE HORSE LATITUDES, or "calms" of Cancer and Capricorn, lie at 30° N and 30° S latitude and are several degrees wide. The winds often drop to calms or occur in erratic breezes that are variable in direction. The skies are usually clear and the atmosphere fresh. The pressures are high but erratic.

According to some, the area got its name of Horse Latitudes because sailing vessels carrying horses across the Atlantic were often becalmed here and, lacking sufficient fodder, had to throw horses overboard.

THE "ROARING FORTIES" is the area of the Westerlies south of the Subtropical High in the Southern Hemisphere where the open ocean dominates. Here the wind blows uninterrupted down an even barometric slope toward the pole, as steadily as the trade winds but with an even greater force. Ships sailed with reduced sail in high seas, hence the name. In the Northern Hemisphere, the barometric pressure on the north side of the Subtropical High decreases unevenly and the winds are not as steady.

HURRICANES—or typhoons as they are called in the western Pacific—are the greatest tropical storms known. Any large circular storm with wind speeds of 75 miles (120 kilometers) per hour is classified as a hurricane. Rotating counterclockwise (in the Northern Hemisphere), they may attain a diameter of over 250 miles (400 kilometers) and a wind speed exceeding 160 miles (260 kilometers) per hour. These great storms may cause severe damage to land areas not only through the force of the wind but also through high tides and drenching rains. North of the equator, they generally occur from July through October. South of the equator, they occur from February through April.

Biloxi, Mississippi, after passage of Hurricane Camille in 1969

Rough seas pound small craft during hurricane.

Hurricanes form in the tropical Atlantic only over the open ocean between 5°-25° latitudes. Their formation requires definite thermal conditions: (1) air and sea temperatures in excess of 77° F. (25° C.), needed to get the air currents started; (2) a slow circular motion, usually at the interface of two air masses, as is found in the Doldrums between the two trade winds system; (3) divergent winds (a high condition) aloft to draw the air out of the center of the circulation, reduce the pressure, and permit the warm air to rise; (4) a distance from the equator

(Number of hurricanes based on records from 1886-1969)

WHEN HURRICANES OCCUR

150

100

50

0

Jan. Feb. Mar. Apr. May June July Aug. Sept. Oct. Nov. Dec.

HURRICANES are powerful storms born at sea, their winds blowing at least 75 miles per hour and commonly 125 to 150 miles per hour. In the center of the giant circular movement, which may be several hundred miles across, is a dead-calm "eye" in which the barometric pressure is at its lowest. The "eye" is 10 to 15 miles in diameter and surrounded by a wall of clouds in which the winds are highest. Forward movement of a hurricane is 10 to 20 miles per hour, with stalls common.

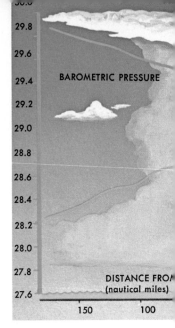

BAROMETRIC PRESSURE

DISTANCE FROM
(nautical miles)

great enough to permit Coriolis force (p. 54) to have effect; and (5) no interferences from land masses.

When these conditions have been fulfilled, the warm air begins to rise through the low pressure center. Air at the surface spirals in to replace the rising air, and the warm air spreads outward from the top of the storm. The force of the winds attained is regulated largely by the temperature of the surface water, decreasing over cool water or increasing and feeding on heat released from such warm surface waters as the Gulf Stream. Over land, with its low specific heat, hurricanes tend to diminish in strength, and the land itself slows the circular

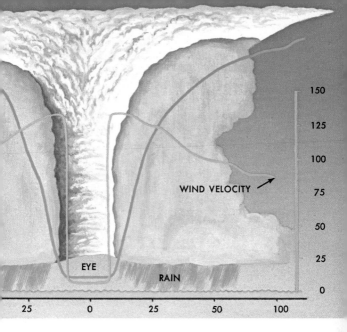

					150
					125
					100
WIND VELOCITY					75
					50
EYE					25
		RAIN			0

| 25 | 0 | 25 | 50 | 100 |

winds. Also, insufficient moisture is obtained from land surface to bolster the energy of latent heat of condensation. The exact structure of the surface winds is not known, though there is evidence that these often consist of jets of strong winds spiralling upward. Most hurricanes are accompanied by heavy rainfall, ranging from 10-20 inches (25-50 cm.) near the storm center.

Hurricanes tend to follow a circular course, moving forward from the equator around a subtropic high. Hurricane prediction and tracking are now greatly facilitated by pictures sent back by Tiros weather satellites, radar tracking, and direct storm observation by specially equipped hurricane hunter planes. Hurricanes are affected by high and low pressure areas in their path, the highs tending to push them away and the lows to draw them in.

Very low barometric readings are found in the center of hurricanes due to the decreases in pressure resulting from the rising air column. This decrease in air pressure may result in excessively high and dangerous tides that flood low coastal areas. Extreme flooding occurs when a large storm with very low pressures moves ashore at the time of a spring high tide.

Photograph of the "eye" of Hurricane Betsy taken on September 2, 1969 near Grand Turk Island, from altitude of about 11 miles.

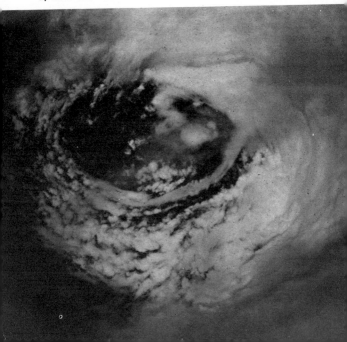

warmer   almost constant temperature   cooler   almost constant temperature

LAND AND SEA BREEZES are encountered in coastal areas and by ships as far as 50 miles (80 kilometers) offshore during periods of fair weather. In the early morning general calms prevail, but by mid-morning light breezes begin to come in from sea. By noon, a strong sea breeze is blowing. In the evening, this sea breeze dies down and is replaced before midnight by a land breeze that persists until morning. These regular shifts of wind direction are due to the different temperatures of land and water. The surface heat of the sea varies only a fraction of a degree between day and night, but the land heat changes significantly.

Under the heat of the morning sun, the land quickly becomes warmer than the sea, and the air over it heats, expands, and rises. This draws in the now cooler air from off the ocean—the sea breeze. During the early evening the land loses its heat and becomes cooler than the sea. Now the air rises from the sea and draws in the cooler air from the land as a land breeze. Land and sea breezes generally do not occur in high latitudes where the differences in land and sea temperatures are not as great as in temperate and tropical regions.

93

The seas are the ancestral home of all life. Many plants and animals today begin life either as eggs or single cells drifting freely in the sea or as embryos developing in watery capsules, still giving evidence of the fluid environment of their ancestors.

Many thousands or even millions of years were probably required for the first low forms of life to develop. During these early years of earth's history, great volcanic activities and earthquakes continually changed the surface of the land. Lakes were formed, then disappeared and were replaced by deserts. Mountains rose and sank again beneath the sea. But through all this time, the oceans remained much the same, regardless of how the land masses were transformed. Only in such a stable environment could life have developed.

Originally sea water contained no oxygen. But primitive organic compounds were formed from inorganic sources in those early seas. Apparently they were simple, naked molecules of DNA (deoxyribonucleic acid) that were able to duplicate themselves. This ability to reproduce is one of the fundamental characteristics of all living things. Later more molecules were added, and finally a simple membrane was constructed. In some such manner, at any rate, the first living organisms developed.

The first primitive forms of life subsisted on the accumulation of complex organic substances. They derived energy from fermentation—converting sugar to alcohol, carbon dioxide, and energy. Eventually the original accumulation of substances in the water was depleted, and mostly those forms of

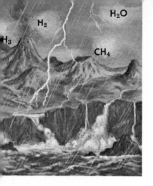

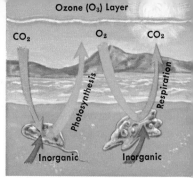

life then survived that could form new protoplasm by photosynthesis. They utilized the carbon dioxide that was by now present in the environment. In this manner, the first green plants came into existence. They released oxygen into the water as a by-product of photosynthesis, thus setting the stage for the development of animal life. Eventually oxygen accumulated in the atmosphere, and animals invaded the land. Slowly, over many millions of years, life evolved to its present great diversity.

Theories such as this are based upon knowledge of the requirements of present living organisms. To survive in fresh water or on land, animals must be able to keep their body salts and fluids at a nearly constant ratio. Protoplasm consists of about 80 percent water, which would soon evaporate on land if it were not enclosed in a protective membrane. This control of body salts and the water content (osmoregulation) is not needed by many marine organisms because the salt content of their body fluids is about the same as in sea water. Except for vertebrates, only brackish water, freshwater, and land organisms need this accessory system.

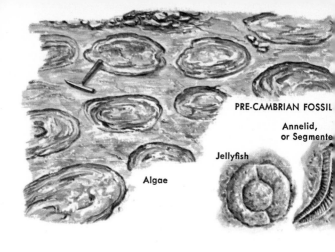

PRE-CAMBRIAN FOSSIL

Annelid,
or Segmented

Jellyfish

Algae

THE FIRST TRACES OF LIFE are found in the sedimentary rocks formed on the bottom of the sea in Pre-Cambrian times. These earliest forms consisted of algae and simple invertebrate animals.

Of the 18 or so major animal groups (phyla), all but two or three are found only in a watery environment, and many are restricted to the oceans, never having invaded fresh water in the millions of years since their evolution.

The present higher forms of life, according to some scientists, may have evolved from a number of simple beginnings, and new primitive forms of life may still be developing today. But today billions of tiny predators in the oceans are ready to feast on helpless forms. The many thousands of years necessary for life to evolve into more complex forms could probably have taken place only in seas devoid of competitive life.

CONDITIONS FOR LIFE IN THE SEA are not as rigorous as for life in fresh water or on land. It has been stated that "blood is modified sea water," and it is true that the salt content of sea water is almost exactly the same as in the body fluids of marine invertebrates. They are isotonic (equal in osmotic pressure; vertebrates are exceptions, but the body fluids of their earliest larval stages are like sea water). Most marine organisms other than vertebrates do not, therefore, need mechanisms to control the salt content of their bodies. But they must be able to adapt when they enter inshore waters of lower salt content. In the open ocean, for example, an amoeba contains no contractile vacuole, but if it is placed in brackish water a contractile vacuole soon appears and begins to pump excess water from its cell.

Echinoderms (starfish and their allies) have no salt-regulating mechanism, hence have never been able to live even in very brackish water. Shore crabs take in and lose water through their mouth and gill membranes. The salts they lose in the process are replaced by salt-secreting cells in their gills. Their kidneys also pick up salts.

Common Sea Hare (Aplysia) becomes distended if placed in water with lower salt content than its body but later assumes near-normal shape as salts in body are removed and equalization occurs.

in natural
sea water

when first put
in brackish
water

much later in
brackish water

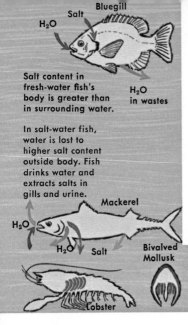

Salt
Bluegill
H_2O

Salt content in fresh-water fish's body is greater than in surrounding water.

H_2O in wastes

In salt-water fish, water is lost to higher salt content outside body. Fish drinks water and extracts salts in gills and urine.

Mackerel

H_2O

H_2O Salt

Bivalved Mollusk

Sea Slug

Fish

Lobster

BODY FLUIDS of bony fishes, which evolved in fresh water and later invaded the oceans, are slightly less salty than sea water (hypotonic), hence water is drawn out of their body toward the greater salt concentration. Because of its tightly fitting scales and coat of slime, a fish's body is nearly impermeable, but water is nevertheless lost through the mouth and gill membranes. A saltwater fish drinks water to make up for this loss. Excess salts are removed by salt-secreting cells in the gills and the salt content of a marine fish's urine is high.

OXYGEN is plentiful in the sea, except under very unusual conditions, and many small and primitive animals get their oxygen by simple transfer of the gas from the sea water into their cells. In larger, more complex forms, oxygen cannot pass into the inner cells by diffusion, and the transfer is accomplished by a circulatory system. Blood is exposed to the sea water and picks up oxygen, which is then carried to body cells. There the oxygen is exchanged for carbon dioxide, which is the process of respiration.

The blood of the most primitive animals was probably simply sea water that contained dissolved oxygen. More active animals required a richer supply of oxygen, and this was supplied by respiratory pigments, such as hemoglobin, dissolved in the blood or concentrated in cells. Some clams and snails pick up oxygen directly through blood vessels in their skin. Others have special gills. Crabs and their relatives pump sea water over gills in a special chamber. Marine mammals have lungs and must come to the surface to breathe.

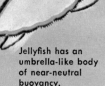

Jellyfish has an umbrella-like body of near-neutral buoyancy.

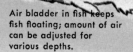

Air bladder in fish keeps fish floating; amount of air can be adjusted for various depths.

Oil droplet in a fish's egg serves as flotation device.

Larva of crustacean has long, much-divided legs which retard sinking.

THE TEMPERATURE OF THE SEA is rather constant—the daily changes are slight, and even seasonal changes are not great. (In contrast, air temperatures may change greatly in short periods.) All marine animals except birds and mammals are cold-blooded—that is, their body temperature is about the same as the surrounding water.

Marine plants and animals are divided into tropical, temperate, or polar groups. Seasonal weather changes are generally not large enough to bother them, but a sudden drop in air temperature to freezing or nearly so in subtropical regions may kill large numbers of fishes in shallow bays and lagoons.

Vertically, marine animals are divided roughly into those of the upper levels of warm water, or thermosphere, with a temperature of about 50° F. (10° C.), and those of cold deep water (psychrosphere).

DENSITY AND VISCOSITY of sea water also influence the distribution of marine animals. Because of its viscosity, sea water is not subject to violent changes, except in the upper levels where it is moved by waves. In the open sea, conditions are rather placid.

Sea water is only slightly less dense than living tissues, so most marine plants and animals are nearly at equilibrium with their surroundings. They do not need to spend as much energy as land animals in counteracting the pull of gravity. Some are suspended by lighter-than-water oil droplets, by gas bladders, or by secreting into their body fluids ions of sodium chloride lighter than those in sea water. Swift-swimming animals are streamlined. Snakelike animals, such as eels, are aided by the density of sea water, which provides a resistance to their body and forces them forward.

99

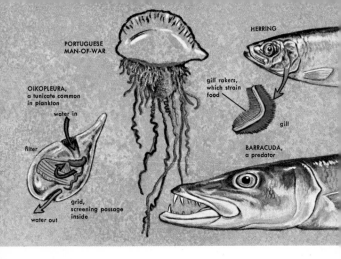

PORTUGUESE MAN-OF-WAR

HERRING

OIKOPLEURA, a tunicate common in plankton

water in

filter

gill rakers, which strain food

gill

water out

grid, screening passage inside

BARRACUDA, a predator

FOOD IS PLENTIFUL in most of the sea, which some biologists have likened to one great rich broth. Sea water examined under a miscroscope reveals great numbers of eggs of invertebrates and fishes, tiny plants, single-celled animals, larvae, and small animals.

Many marine animals feed by filtering food from this rich supply. Tunicates, clams, and other marine animals pump water through their body cavity and strain food from it on the latticework of their branchial baskets, or gills. Others, such as some shrimplike animals, swim forward and strain water through hairs on their mouthparts, licking the food from them at intervals. Some filter-feeders take in food in this manner whether they wish

to or not and pass it out completely undigested.

Jellyfishes, the Portuguese-man-o'war, and others catch their food by hanging barriers of tentacles into the water. With their gill-rakers, some fish strain food from water passing over their gills. Except in spiders and their webs, filter-feeding has never been developed in land animals.

Predators catch their prey by means of swift speed and large teeth, or they trap them with lures, shock them with electricity, or poison them with little darts. For this reason protective devices are highly developed in marine animals. Many are camouflaged by their color or shape. Others are so spiny, repulsively flavored, or poisonous that they are avoided.

REPRODUCTION in single-celled plants, such as the diatoms, and in the single-celled animals, such as the amoeba, is by simple cell division. Higher animals, such as the whales, reproduce by internal fertilization and a long internal development of the embryo.

Due to the similarity of sea water to body fluids, the most common means of sexual reproduction is the spewing forth of sperm into the open sea, where they meet and fertilize the egg. The embryo develops and hatches in the sea. This appears at first to be an extremely wasteful method, with millions of unfertilized eggs drifting off to sink to the bottom, but practically 100 percent fertilization occurs. Unfertilized eggs eventually are either eaten or decompose and aid in recycling nutrients.

In simple spawners, there is usually some kind of trigger mechanism to insure release of sperm and eggs at the proper time. When the weather warms in the spring, for example, the oysters' gonads are stimulated. As soon as a few oysters in a bed begin to release eggs and sperm, a chemical secretion is released that spreads and activates all of the oysters.

Spawning fish commonly swim in schools or in a circle, causing an eddy into which the eggs and sperm are drawn.

Generally, the less parental care offered to eggs and young, the greater the number of eggs produced. Oysters, for example, release millions of eggs and sperm. Sharks give birth to living young and produce only a few. Between these two extremes, protection ranges from carrying eggs on swimmerets (swimming appendages of arthropods), communal egg masses attached to the bottom, or enclosure of the eggs in horny cases.

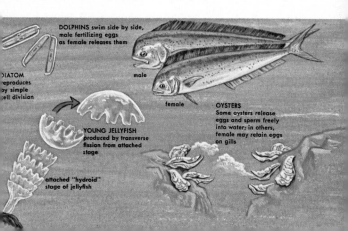

DOLPHINS swim side by side, male fertilizing eggs as female releases them

DIATOM reproduces by simple cell division

male

female

YOUNG JELLYFISH produced by transverse fission from attached stage

attached "hydroid" stage of jellyfish

OYSTERS Some oysters release eggs and sperm freely into water; in others, female may retain eggs on gills

THE DIVERSITY OF LIFE in the sea ranges from plants and animals so small they can be seen only with high-powered microscopes to the gigantic Blue Whale, the largest form of life that has ever existed. The plants and animals comprising the plankton are by far the most abundant, and all are so small that they are carried about by the currents.

Fishes, squids, sharks, porpoises, and whales are nektonic animals. They can swim well and direct their courses even in strong currents. Plants and animals of the seashore and the animals of the bottoms of the seas form the benthos. Many animals of the shore and ocean bottom—the crabs, crawfish, lobsters, clams, snails, and others—either cannot swim at all or swim poorly. The plants that grow attached to the bottom include the common seaweeds, which are large algae. They grow from the shore to the depth of light penetration.

PLANKTONIC PLANTS AND ANIMALS live in the oceans in incredible multitudes. By function, they are divided into three categories: autotrophs, heterotrophs, and phagotrophs. Autotrophs live on inorganic materials; with the aid of sunlight or chemical energy they utilize carbon in the form of carbon dioxide and manufacture food by photosynthesis. Phagotrophs obtain carbon by breaking down organic material. Heterotrophs occur in the lighted zones of all oceans throughout the world. They are by far the most important marine microorganisms, constituting about 98 percent of all the plant life in the sea. The flagellates among them swim, using their whiplike flagella. Others retard sinking by having processes projecting from their shells,

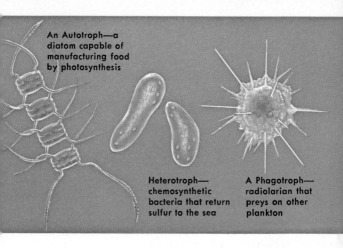

An Autotroph—a diatom capable of manufacturing food by photosynthesis

Heterotroph—chemosynthetic bacteria that return sulfur to the sea

A Phagotroph—radiolarian that preys on other plankton

and still others contain oil droplets that assist them in floating in the upper sunlit areas (pp. 104-105). Diatoms are phytoplankton that swarm in cold seas, often discoloring the water. Blue-green algae, coccolithophores, and others are abundant in warm and tropical waters. Dinoflagellates may turn the water red or brown and they give off neurotoxins, poisonous to fish and to invertebrates.

The most important heterotrophs are the bacteria that swarm in the sea from the surface to the greatest depths. They decompose plant and animal remains, thus returning sulfur, nitrogen, phosphate, carbon dioxide, and other essentials to the sea.

Foraminifera, radiolaria, and other minute protozoans feed phagotrophically—taking in bacteria, phytoplankton, and detritus. The calcareous shells of foraminifera and the siliceous shells of diatoms and radiolarians form characteristic oozes (p. 37).

DIATOMS

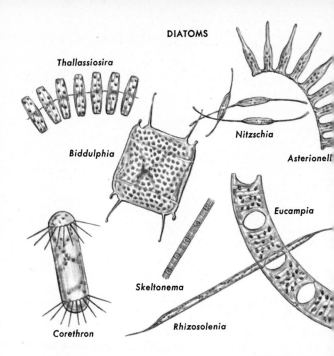

Thallassiosira

Biddulphia

Nitzschia

Asterionell

Eucampia

Corethron

Skeltonema

Rhizosolenia

PHYTOPLANKTON are mostly inactive, especially the diatoms, but some can swim.

Diatoms, especially common in cool waters, are the most important group of phytoplankton. They may occur singly or in long chains of many individuals. Each plant, the largest of which is barely visible to the naked eye, is enclosed in a thin siliceous shell composed of two parts that fit one inside the other. The shells may be sculptured with lines, pits, or nodules.

Diatoms reproduce by simple division, or fission. Each new plant receives only one part of the parent shell, hence some diatoms get smaller and smaller while others remain about the same size. After a few generations of reducing, the small diatoms slip out of their glasslike casing and into an expandable membrane. They grow to full size before again acquiring a shell. Many diatoms go into a resting spore stage during winter or in other unfavorable periods.

DINOFLAGELLATES

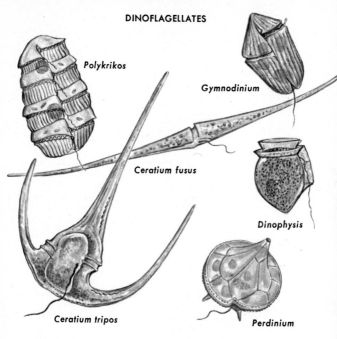

Polykrikos

Gymnodinium

Ceratium fusus

Dinophysis

Ceratium tripos

Perdinium

Dinoflagellates are the second most important group of the phytoplankton. Most of them are capable of photosynthesis, others feed like animals, and some live on decomposng organic matter. Some dinoflagellate cells are naked; others are covered with an armorlike plate of cellulose. All, however, are equipped with whiplike structures, or flagella, by which they move about. Dinoflagellates vary in shape from little disks to bent needles. They reproduce by fission.

All phytoplankton require sunlight to produce food. They are equipped with various structures to keep themselves in the upper layers of the sea. The shells of many have filaments or other "flotation processes" that increase the cell's surface and thus retard sinking. Flotation is accomplished mainly by tiny oil droplets.

Silicoflagellates, coccolithophores, and heterococcales are other phytoplankton. Some are only 5-20 microns in diameter, the smallest plants in the sea.

105

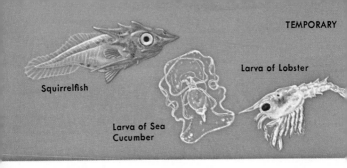

Squirrelfish

Larva of Sea Cucumber

Larva of Lobster

ZOOPLANKTON are diversely shaped, weird little animals with representatives in every major group of marine animals. Some are temporary plankton. These consist of larvae of animals that as adults belong either to the benthos or the nekton. Permanent plankton spend their complete life cycles as plankton.

In the temporary plankton group are the larvae of sponges, corals, worms, mollusks, echinoderms, fishes, and animals in many smaller, lesser known groups. In many, the eggs are shed into the sea to float about until hatching.

These planktonic creatures must stay in the upper layers of the sea to obtain food. Some feed on phytoplankton; others are carnivorous. But directly or indirectly, all depend on the phytoplankton. Like the phytoplankton, these small animals may be equipped with long spines, bristly appendages, or be flattened—all aids in keeping them afloat.

In the temporary planktonic stage the young of a species are supplied copiously with food in the ocean currents, which also distribute the species over a wider area than would otherwise be possible.

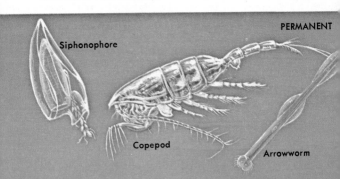

Siphonophore

Copepod

Arrowworm

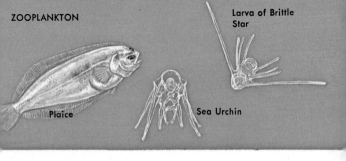

ZOOPLANKTON

Larva of Brittle Star

Plaice

Sea Urchin

If the larvae are in the proper depth of water or over the right type of bottom when they change into juvenile form and are ready to begin living like adults, they survive. If not, they die and decompose, thus returning to the food cycle. Some temporary plankton that do not find a suitable place to settle can prolong their larval life almost indefinitely.

Permanent plankton consist of free-living protozoans (foraminifera, radiolarians and tintinnids). A few types of jellyfishes spend their adult life drifting in the sea, as do siphonophores, ctenophores, some worms, rotifers, chaetognaths, such crustaceans as some kinds of shrimps, a few crabs, ostracods, copepods, mysids, cumacids, euphausiids, amphipods and others, heteropods and pteropods among the mollusks, a few echinoderms, and a number of planktonic tunicates. The herbivorous members of this group are filter-feeders, straining diatoms and dinoflagellates from the water. They may be fed on by carnivorous animals. Permanent plankton occur in tremendous numbers in the sea and are the principal food of many large fishes and even whales.

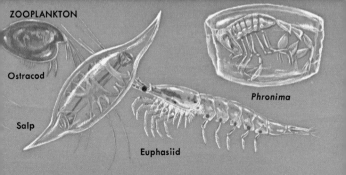

ZOOPLANKTON

Ostracod

Salp

Euphasiid

Phronima

NEKTON include all the large swimming animals, which are mainly the squids, fishes, and marine mammals. Most are predators, searching actively for their food and seizing it with teeth or with other specialized catching apparatus. Fewer kinds strain plankton from the sea—the fishes by means of rakers on their gills, the whales with their bony sieve of baleen.

Nektonic animals are most numerous in surface waters, where food is very abundant. They decrease in numbers with depth. But little is known about many nektonic animals, either their distribution or their numbers. Squids, for example, are difficult to capture with nets, and something of what we know about their numbers has been learned through finding them in the stomachs of Sperm Whales and other toothed whales that feed principally on squids.

Nekton are not evenly distributed in the ocean. They depend for their food on the zooplankton that are found in great numbers in areas of up-

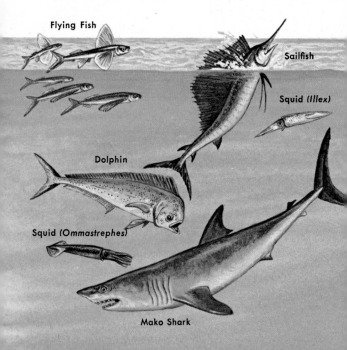

Flying Fish

Sailfish

Squid (Illex)

Dolphin

Squid (Ommastrephes)

Mako Shark

welling where nutrient-rich waters are brought back to the surface. They are abundant, too, along the edges of ocean currents and where currents divide; these are places where the movement of the water forms turbulence and permits mixing of the different water layers.

Places with no upwellings and currents contain comparatively few living things; they are like the deserts on land. The Sargasso Sea, for instance, is too far from land to derive nutrients in the runoff from rivers, and its waters are practically undisturbed. Very few large fishes are found here. In contrast, concentrations of tunas, marlins, and other large fishes may be found at the edges of the North and South Equatorial currents, even in mid-ocean.

Few areas of the sea contain the great quantity of plankton required by the filter-feeding whales. These animals are concentrated in polar waters, where the plankton occur abundantly in the shallow zone of light penetration.

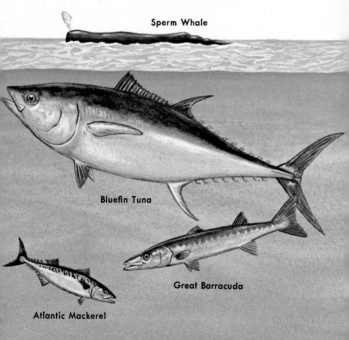

Sperm Whale

Bluefin Tuna

Great Barracuda

Atlantic Mackerel

PLANTS ARE THE BASIS of life in the sea exactly as they are on land. They utilize nutrients dissolved in sea water, hence they can live anywhere in the sea where there is sufficient light for photosynthesis. Phytoplankton, the tiny single-celled plants, swarm in the open sea in a layer of water from two or three feet (.6 to 1 meter) deep in polar seas to several hundred feet (50 to 100 meters) deep in warm and tropic seas. They account for about 80 percent of the total photosynthesis on earth. Food production may be fifty times greater along shores than in the open sea, but shallow shore waters constitute only a fraction of the total volume of the oceans.

In its simplest form, the equation for photosynthesis is $6CO_2 + 6H_2O \rightarrow C_6H_{12}O_6 + 6O_2$. Carbon dioxide is always present in large quantities in the sea. There is no lack of water, and the plants contain the green pigment, chlorophyll, that triggers the process. Sunlight is the limiting factor. If a plant sinks too deep or if the sunlight becomes too weak, the plant may not have sufficient oxygen, a by-product of photosynthesis, for its own respiration. Most of the dissolved oxygen results from photosynthesis.

PHOTOSYNTHESIS furnishes the carbohydrates necessary for life. Nitrogen and phosphorus, in the form of nitrites-nitrates and phosphates, are also needed, and these chemicals can be used directly by plants. They occur throughout the oceans but are found in largest quantities in near-shore waters and in areas of upwellings (p. 73)

Nitrites-nitrates and phosphates are released from organic matter by bacterial decomposition. Since they occur in limited amounts but are essential to plant growth, they are limiting factors in plant distribution. Trace elements are necessary in minute amounts. When their needs are present, marine plants grow and reproduce at astounding rates.

THE COMPENSATION DEPTH is the level in the sea where the light intensity is just sufficient for oxygen production to balance oxygen consumption. Plants sinking below this level will soon die unless brought back by upwellings into the euphotic zone, the layer above the compensation depth.

IN NORTHERN SEAS when the sun rises sufficiently high for light to penetrate strongly enough to permit photosynthesis, diatoms and dinoflagellates become so abundant that they color the sea. The compensation depth may be only a few meters off New England or in the North Sea. In warm seas, the compensation depth may be at 100 m or more.

A few weeks after conditions have become favorable, the "bloom" reaches its height. Then, as phosphates and nitrates are used up, the phytoplankton diminish sharply in numbers. By midsummer most of the plants of the first "bloom" have been either eaten or are decomposed by bacteria. Phosphates and nitrates are again available, and a second or "fall bloom" takes place. It is not as large as the "spring bloom," and the nutrients are depleted more rapidly.

In cold weather, most phytoplankton go into a resting stage, and productivity in general stops. Bacteria continue to be active during the winter, subsisting on nutrients brought to the surface from below the thermocline (p. 46) by winter storm seas. They set the stage for the cycle to begin again in spring.

In warm seas, fluctuations are very slight, since seasonal changes are not as great. Phytoplankton exist at a lower but more constant level all year.

Cyclic Production of Diatoms in North Sea

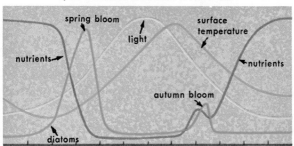

Jan. Feb. Mar. Apr. May June July Aug. Sept. Oct. Nov. Dec.

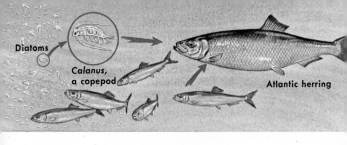

Diatoms

Calanus, a copepod

Atlantic herring

FOOD CHAINS IN THE SEA may be simple or complex. Mineral salts are derived basically from the land through erosion and freshwater runoff and secondarily from decomposed organic material. Oxygen and carbon dioxide are supplied mainly from the air and secondarily from animal respiration and plant photosynthesis. All of these, plus light energy from the sun, are essential for the growth of plants.

Some plants die and are reduced by bacteria to their basic nitrogen and phosphorus compounds and trace elements, ready for use again by new generations of plants. Another step is added if a filter-feeding herbivore, such as the copepod *Calanus*, feeds on diatoms. If the copepod dies and

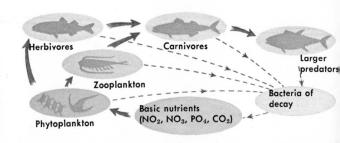

Herbivores

Carnivores

Larger predators

Zooplankton

Bacteria of decay

Phytoplankton

Basic nutrients (NO_2, NO_3, PO_4, CO_2)

Man

decomposes, the basic chemicals are again put in circulation. A third step is introduced if a carnivore —probably the North Atlantic herring in this case— feeds on the *Calanus*. The herring may be eaten by a larger fish or by man, or it may die of natural causes. Many steps are possible before the nutrients are returned to the sea.

Sometimes, food chains are drastically altered, as happened along the islands off Peru and Ecuador. The food chain there starts off like others, but at the level of a filter-feeding carnivorous sardine, sea birds enter the cycle and feed on the fish. They return little to the sea directly, for their droppings are concentrated on land at certain nesting and perching sites. These great accumulations form the guano deposits that are an important source of phosphates and nitrates in fertilizers. These coastal South American countries have found that the sardines have great commercial possibilities in the form of oils and meal, but if the sardines are removed in tremendous numbers, the guano industry suffers and may be seriously depleted. These governments are faced with a major decision—whether to break the food chain link or to allow nature to continue as it has for millions of years.

113

LIFE ZONES IN THE OCEAN are of two kinds: (1) horizontal, based on the plants and animals that live in the surface layers of tropical, temperate, and polar regions; and (2) vertical, which extends from the surface to the greatest depths.

Distribution of life in the sea, however, is limited also by light penetration and by pressure, and so organisms living at the surface in polar regions can become distributed from one side of the tropical region to the other only if they can tolerate vast changes in pressure and do not need sunlight. Plants or animals with broad temperature and pressure ranges are called eurythermal and eurybathic; those with limited temperature and pressure tolerances are called stenothermal and stenobathic.

HORIZONTAL DISTRIBUTION varies with temperature. Tropical oceans are those in which all of the surface waters are 68.5° F. (20° C.) or above. The tropical zone is widest on the western sides of the oceans, where the major tropical ocean currents divide and go toward the poles. On either side of the tropical region are the temperate regions, bounded by the 68.5° F. isotherm toward the equator and by the 50.5° F. (10° C.) isotherm toward the poles. Animals and plants in the warmer part of the temperate zone are called warm-temperate organisms; those on the polar sides are called cold-temperate. The polar regions are marked by the 10° C. isotherm where the surface waters seldom become warmer than 10° Centigrade.

VERTICAL DISTRIBUTION of marine animals is also evident. The tropical regions are limited to a fairly shallow layer—about 1,000 to 1,300 feet (300 to 400 meters) deep. This region is cradled in temperate zone waters that extend from the surface isotherm of 50° F. (10° C.) in northern seas completely across and beneath the warm waters of the tropical region to the 10° C. isotherm at the surface in the Southern Hemisphere. Likewise, the polar waters dip beneath both the temperate and tropical waters, extending from the Arctic Ocean to the surface in the Southern Ocean. Life found in either polar region might be distributed through the cold waters from one pole to the other, but in the equatorial region it would live below 2,300 feet (700 meters).

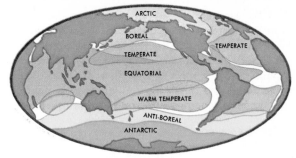

Generalized distribution of the major types of plankton

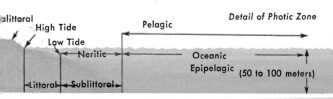

Detail of Photic Zone

alittoral

High Tide

Low Tide

Pelagic

Neritic

Littoral | Sublittoral

Oceanic

Epipelagic (50 to 100 meters)

Bands of color indicate areas in which same types of plankton occur.

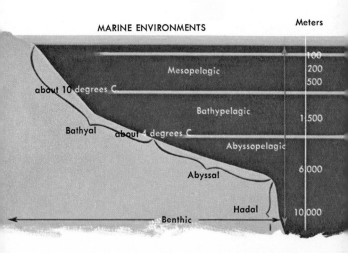

MARINE ENVIRONMENTS

Meters

Mesopelagic

100
200
500

about 10 degrees C

Bathypelagic

1,500

Bathyal

about 4 degrees C

Abyssopelagic

6,000

Abyssal

Hadal

10,000

Benthic

PELAGIC plants and animals live in vertical zones from the surface to the deep sea. They are surface dwellers (epipelagic), middle (mesopelagic), deep-sea (bathypelagic), or abyssal (abyssopelagic or hadopelagic). Benthic forms live on the bottom. These zones are generally limited by the depth of light penetration, pressure, and temperature.

THE EPIPELAGIC ZONE is the lighted zone of the sea, where phytoplankton and the host of zooplankton and larger organisms that feed on them flourish. The greatest amount of life in the open seas occurs in this layer. In the higher latitudes, the epipelagic zone may be only a few feet (meters) deep; in tropical and temperate areas, it may be less than

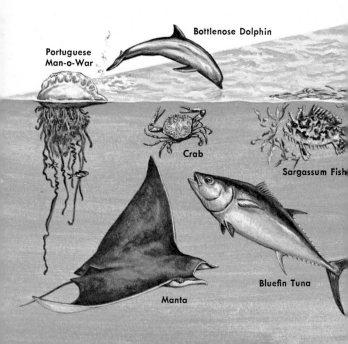

Bottlenose Dolphin

Portuguese Man-o-War

Crab

Sargassum Fish

Manta

Bluefin Tuna

300 feet (100 meters) deep on the eastern sides of oceans but 650 to 1,000 feet (200 to 300 meters) deep in clear subtropical waters on the western sides.

Animals of the epipelagic zone are well adapted to this near-surface habitat. Fishes have silvery undersides, making them almost invisible to a predator looking upward from beneath toward the light on the surface. Their backs are a deep indigo blue or purple, which obscures them from the sharp eyes of predators looking down into the dark depths.

Many epipelagic fishes are very fast swimmers. Others have special adaptations for escape. Flying fishes can surge from beneath the water and glide several feet above the surface for 100 to 500 feet. Sargassum fish resemble the Sargasso weed in which they live, camouflaged by their fringed fins and blotched colors. Zooplankton found in the epipelagic zone are often colorless and thus quite invisible, or they are tinged with blue or green that blends them with the sky as viewed from below.

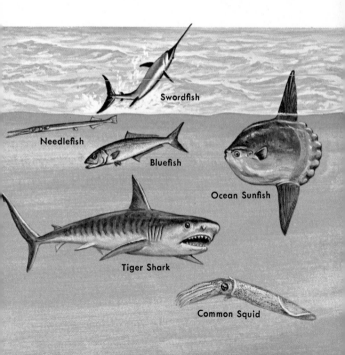

Swordfish

Needlefish

Bluefish

Ocean Sunfish

Tiger Shark

Common Squid

THE MESOPELAGIC OR TWILIGHT ZONE lies below the epipelagic zone and receives little light from above. It is inhabited by darkness-loving (scotophilic) animals. Their upward movements are limited by the penetration of sunlight, their lower movements by the ocean's cold deep layers—the 50.5° F. (10° C.) isotherm at about 1,600 to 3,200 feet (700 to 1,000 meters). This same temperature line marks the limits of the subpolar regions.

Animals of the mesopelagic zone often migrate to the surface at night, then return to the dark depths when daylight comes. Since they live below the true photic zone where primary production takes place, food sources are not as great as for those that live in the zone above. But because they forage in the upper zone at night, their numbers are still rather high.

Animals of the mesopelagic zone have firm bodies, commonly somewhat silvery beneath and very dark above. Their eyes are larger than the eyes of their surface-dwelling relatives, permitting them to see even in the dim light.

Many mesopelagic animals possess light organs, producing a cold or biological light similar to that of fireflies. The chemical involved is luciferin, which in the presence of luciferase and oxygen emits a beautiful pale glow. Some of these light organs, especially in some species of the squids, are

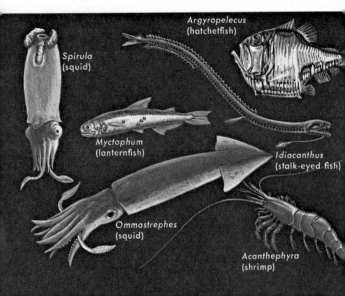

Spirula (squid)

Argyropelecus (hatchetfish)

Myctophum (lanternfish)

Idiacanthus (stalk-eyed fish)

Ommastrephes (squid)

Acanthephyra (shrimp)

complex, consisting of a light source (luciferin-luciferase) contained above a layer of reflector cells. Muscles can move the whole unit toward or away from a lens for focusing, and a diaphragm near the lens can be contracted to close off the light. The whole structure may be surrounded by chromatophores (color cells) that can be expanded in different color combinations and thus change the color of the light being emitted by the animals.

Bioluminescence may serve to keep schools of animals together or to separate species. It probably functions also in distinguishing or attracting opposite sexes. Some may employ it to scare away predators, others to attract plankton.

Like headlights, these light organs may illuminate the way for some animals. *Diaphus*, a small fish, has one very large light organ above each eye. Its flashing would momentarily blind the fish when turned on in the dark depths, but another very tiny light organ associated more closely with the eye is turned on first. Thus, the fish's eye is adjusted to this increased brightness when the larger lights come on a second later.

Common animals of the mesopelagic zone are the lantern fishes, hatchet fishes, bright red shrimps, and numerous unusual squids, such as *Spirula* and the fire squids. Many that live in the deeper parts of the zone are bright red or black.

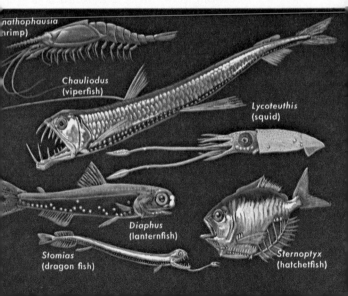

Gnathophausia (shrimp)

Chauliodus (viperfish)

Lycoteuthis (squid)

Diaphus (lanternfish)

Stomias (dragon fish)

Sternoptyx (hatchetfish)

THE BATHYPELAGIC ZONE, beginning at an average depth of about 2,300 feet (700 meters), is characterized mostly by waters as cold as in polar regions. The true bathypelagic regions extend from approximately 2,300 feet (700 meters), where the temperature drops abruptly, down to about 6,500 feet (2,000 meters), where the water temperature is only slightly above 40° F. (4° C.). This cold psychrosphere occupies most of the oceans.

Animals of the bathypelagic zone live in perpetual darkness. Generally, they are soft-bodied and somewhat gelatinous, their outer skin so thin that it rubs off at the slightest touch. They are usually a sooty black or a deep red. Bathypelagic animals live considerably below the zone of food production, hence predation is highly developed. Bioluminescence is reduced; most animals have small eyes, and some are blind.

Among the weird animals from these depths are the deep-sea angler fishes, with luminescent lures dangling in front of their large mouths. Viperfishes have tremendously long and numerous saberlike teeth, and many have rows of lights along their sides. Squids are delicate, elongated creatures, often with long, stalked eyes and bodies so transparent that their internal organs may be plainly seen. Many animals have an extensible stomach that permits them to digest animals larger than themselves.

In these dark depths, finding a mate is not easy. In some of the angler fishes, the young male is free-swimming, but when he contacts a female he attaches to her side with his teeth. Soon he metamorphoses to a small sac with only a fin-like object sticking out. Then the male derives nourishment from her circulatory system. Some females may carry 7 or 8 males.

THE ABYSSOPELAGIC ZONE, separated from the bathypelagic zone by the 40° F. (4° C.) isotherm, extends downward toward the greatest depths of the sea and ends just above the bottom. Little is known about life in this great depth. The zone contains few animals —simple crustaceans, small angler fishes, radiolarians, and a few other organisms—widely scattered due to the lack of food. Even less is known about the pelagic life below 20,000 feet (6,000 meters). This is the hadopelagic region, found only in the waters lying above the bottom of the very deepest trenches in the oceans. Temperatures here may range to as low as 29° F. (—2° C.), and the pressures are as great as 7 tons per square inch. Food is so scarce in these depths that only a very small population can be supported. Probably fewer than a dozen hauls have been taken in the waters of the great trenches.

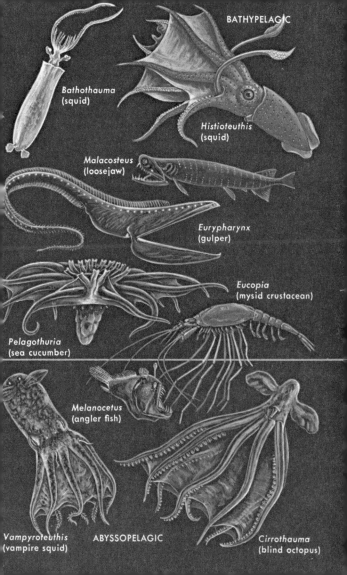

BATHYPELAGIC

Bathothauma
(squid)

Histioteuthis
(squid)

Malacosteus
(loosejaw)

Eurypharynx
(gulper)

Eucopia
(mysid crustacean)

Pelagothuria
(sea cucumber)

Melanocetus
(angler fish)

Vampyroteuthis
(vampire squid)   ABYSSOPELAGIC

Cirrothauma
(blind octopus)

THE BENTHOS, or life on the bottom, is as varied as the pelagic life (p. 116) and also occurs in zones from the shore to the depths. Some areas are rich; others are veritable deserts. Like pelagic forms, bottom-dwelling organisms are limited in distribution by temperature, and each kind of sediment or bottom (mud, sand, rock) harbors a characteristic plant and animal group. Beyond the edge of the continental shelf, muds and oozes differ mainly in whether the shells of the organisms forming them were made of lime or of silica.

Food may be abundant or scarce on the bottom. In shallow waters, of course, food may be very abundant due to the heavy concentrations of plants and animals in the surface waters directly above. In the depths, the bottom-dwelling animals depend on plant or animal remains that drift down from above. This settling of organic remains may be less than 0.5 mm. per year (p. 37).

BALANCED BENTHIC COMMUNITIES were discovered by European biologists a few years ago. Whether in arctic or tropical waters, the same general types of marine invertebrates are usually found in the muds and sands, often in about the same numbers.

If a particular kind is missing from the community, it can be introduced and will survive, because a natural niche has been left open. In Danish waters, a community was found that should normally have had soft-shelled clams, but none were present. The Danish government imported live soft-shelled clams, and in a few years the thriving clam beds formed a new fishery.

The biological mechanism by which these communities are able to maintain themselves was also an interesting discovery. When the young of some species are settling to the bottom, they are likely prey for the adults of other species. But at this same time, the potential predators are unable to feed because they are about to bear young. By the time they are again ready to feed, the young of the other species are too large to be picked up in their filtering systems.

THE OUTER SUBLITTORAL ZONE, covering only about 5 percent of the world's oceans, is the area of the continental shelf from about 150 feet (50 meters) deep to about 600 feet (200 meters)—or to the outer edge of the shelf. Most of this water is below the zone of light penetration, hence below the area of primary productivity, yet this region is one of the richest in the world in animal life. About 90 percent of all the fish taken commercially come from this zone, which includes the great fishing grounds of Georges Bank, the Grand Banks, and the North Sea. Except at high latitudes, the temperature of the water is higher than 50° F. (10° C.).

Where bottom sediments from the land meet the mud zone that extends down into the deeper parts of the oceans, the outer sublittoral zone is particularly rich. Sir John Murray called this area of high organic content the "mudline." Snappers, groupers, hakes, haddocks, cod, flounders, and sole are among the common mudline fishes. Most common of the many invertebrates are large starfishes, serpent stars, sea anemones, sea cucumbers, crabs, lobsters, and burrowing worms. Nearly every major group in the animal kingdom is well represented. Unfortunately, only a few of the multitude of species found here are now considered edible.

Fish catch from middle shelf region off Colombia.

Otter trawl haul from about 200 meters off coast of West Africa.

THE BATHYBENTHIC ZONE extends from the edge of the continental shelf down the continental slope and onto the ocean floor to a depth of about 13,000 feet (4,000 meters). Thus, it parallels the mesopelagic and bathypelagic zones of the open ocean. The water temperature varies from about 50° F. (10° C.) to 40°F. (4° C.).

Not a great deal is known about life communities on the continental slope. Because the slope is fairly steep, it is difficult to keep trawls and dredges on the bottom, or they are caught up and lost in the gorges, canyons, valleys, and sharp peaks that occur on the rough bottom.

Animal life in this zone becomes less dense with greater depth, and typical bottom life is encountered at about 6,500 feet (2,000 meters). Here the bottom is generally smooth, corresponding to what was formerly called the abyssal plain. Except for occasional hills and ridges, trawls can be dragged across the bottom.

The crustaceans that live in the bathybenthic zone are primarily deep red, though some are very pale. They have long legs, an adaptation for support on the soft bottom, and in most the eyes are reduced in size. Bottom-dwelling fishes of this zone are smaller than those in the waters above. Many of them have long feelers, spines, armor, and other unusual adaptations for procuring food or for supporting themselves in the ooze.

124

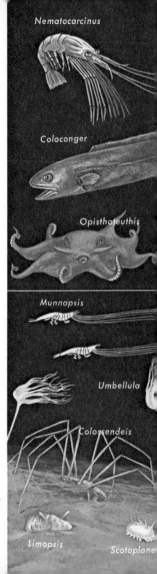

Nematocarcinus

Coloconger

Opisthoteuthis

Munnopsis

Umbellula

Colossendeis

Limopsis

Scotoplane

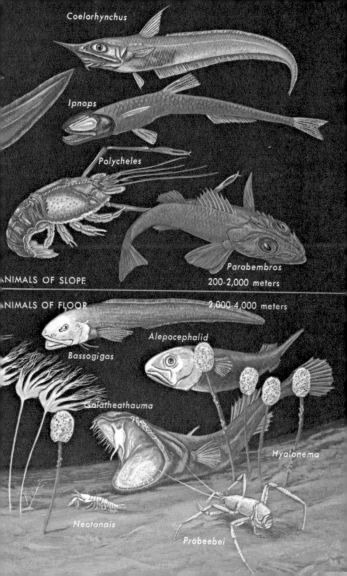

Coelorhynchus

Ipnops

Polycheles

Parabembros

ANIMALS OF SLOPE
200-2,000 meters

ANIMALS OF FLOOR
2,000-4,000 meters

Bassogigas

Alepocephalid

Galatheathauma

Hyalonema

Neotanais

Probeebei

ABYSSOBENTHIC animals live below 13,000 feet (4,000 meters), where the temperature is always less than 40° F. (4° C.). The bottom is soft, composed of various oozes formed of the dead skeletons of planktonic organisms dropped from the seas far above. Food consists mainly of organic debris and bacteria. Those few kinds of animals that can subsist on this meager diet are small crustaceans (amphipods and some unusual shrimplike creatures), radiolarians, protozoans, soft sea anemones, sea pens that luminesce on stalks projecting above the bottom muds, flabby sea cucumbers, tooth shells, and serpent stars.

The few fishes that live in these deep dark waters are mostly colorless; many are blind. Some kinds walk about the bottom on long, slender, stiltlike fins that raise them above the oozes.

Trawls in these depths bring up manganese nodules (concretions of this mineral formed around various hard cores). Whale earbones and sharks' teeth, too hard to dissolve as they drift down from above, are also found in the oozes.

THE HADOBENTHIC ZONE extends from 20,000 feet (6,000 meters) down to the greatest depths known in the oceans (over 36,000 feet or 10,800 meters in the deep trenches). The trenches are long, slender, and steep-sided, their narrow flat bottoms consisting of soft sediments on which it is possible to trawl for samples.

Little was known about the animals of these depths until 1948, when the Swedish Deepsea Expedition obtained specimens from a depth of over 23,000 feet (7,000 meters) in the Puerto Rico Trench in the western Atlantic. Many hauls have been made in great depths since then by scientists of the Danish Deepsea Expedition from the *Galathea* and by American and Soviet scientists. The temperature is from 32° F. to 29° F. (0° C. to —2° C.), and the pressure exceeds 10 tons per square inch. Except for detrital material and perhaps large populations of bacteria, food is extremely scarce. Yet the animal population is high. In only one haul, the *Galathea* brought up over 3,000 specimens of a small sea cucumber.

Most forms of life in these depths are very small, consisting of holothurians or sea cucumbers, polychaete worms, sea lilies, starfishes, brittle stars, mollusks, and a few small crustaceans. Only two or three fishes have been reported from below 20,000 feet (6,000 meters). Because many of the animals appear to be bioluminescent, it is believed that the bottom is lighted by a soft luminescent glow. Each trench seems to harbor a unique group of animals, much as islands and continents do.

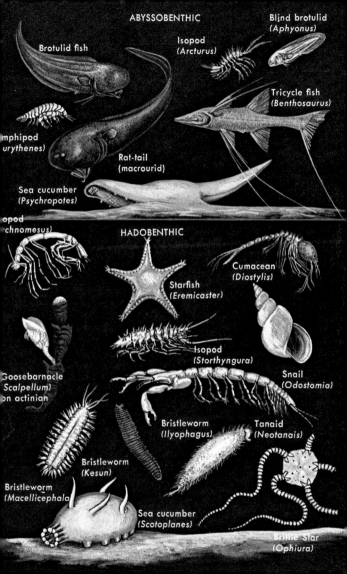

ABYSSOBENTHIC

Brotulid fish

Isopod
(Arcturus)

Blind brotulid
(Aphyonus)

Tricycle fish
(Benthosaurus)

mphipod
urythenes)

Rat-tail
(macrourid)

Sea cucumber
(Psychropotes)

opod
chnomesus)

HADOBENTHIC

Starfish
(Eremicaster)

Cumacean
(Diostylis)

Isopod
(Storthyngura)

Goosebarnacle
Scalpellum)
on actinian

Snail
(Odostomia)

Bristleworm
(Ilyophagus)

Tanaid
(Neotanais)

Bristleworm
(Kesun)

Bristleworm
(Macellicephala)

Sea cucumber
(Scotoplanes)

Brittle Star
(Ophiura)

FISHERIES OCEANOGRAPHY

Fisheries oceanographers are concerned with oceanographic factors influencing or controlling the distribution and numbers of economically important organisms. By determining these factors—temperature, salinity, nutrients, trace elements, upwellings, and many others—they are able to assist in the location of new stocks of harvestable organisms.

One of the largest surveys of fisheries was Equalant (Equatorial Atlantic), in which multivessel studies were conducted synchronously from the Caribbean Sea and Brazil to western Africa.

Fisheries oceanographers search for upwellings or areas where currents running alongside each other create turbulence that breaks up the thermocline and permits nutrients to reach the surface, providing enrichment for productivity in the photic zone. By sampling the plankton systematically over broad areas, they can locate zones of high productivity or barren areas, thus helping to predict where various types of fisheries may flourish. By chemical analysis of sea-water samples from various depths and by illumination studies, they may determine the depth of nutrient-rich waters and whether there is sufficient light for photosynthesis of phytoplankton. Of increasing interest today are studies of pollution and its effect upon important and valuable fisheries.

Islands and underwater rises break up the thermocline creating areas rich with food.

Biologists measure mixed catch taken aboard *Albatross IV*, a research vessel operated by National Marine Fisheries Service.

GREATER USE OF FISHERIES RESOURCES is necessary to feed the increased world population—now more than 3½ billion. By the Western World standard that each person should have about 70 grams of protein per day for good health, approximately 100 million tons of protein are needed per year to feed the earth's population now. At present, approximately 67 percent of this protein comes from plants, 24 percent from meat and milk, 4 percent from poultry and eggs, and 5 percent from fish.

This standard is far from being met on a world-wide basis. Only about 19.5 percent of humans average more than 30 grams of animal protein in their daily diet. About 19.8 per-cent have about 15 to 30 grams per day, and about 60.7 percent less than 15 grams per day. Thus, the majority of the world's people today have a protein deficiency.

In Japan, the average daily protein diet is only 15 grams; in Egypt, 13 grams; in Pakistan, 8 grams; and in India, 6 grams. These low-protein countries are concentrated in the rapidly expanding and developing tropical and subtropical regions.

The total protein yield from the sea accounts for only about 10 percent of the world food supply at present. Unless yields from the land can be increased greatly to feed the world's growing population, the seas must be turned to even more to satisfy human needs.

129

MARINE HARVESTS were nearly all taken off the continental shelf until the early 1950's. Less than 10 percent of the ocean surface was utilized, and the vast remainder was untouched. Fisheries oceanographers were at that time concerned almost wholly with the complex chemistry and hydrography of shallow waters, primarily with coastal currents and upwellings; with the enriching effect of water run-off from bays, harbors, and rivers; and with the turbulence of oceanic waves in returning nutrients from the shelf. They also studied life histories and the various links in food chains that might affect them. They assembled statistics of the numbers of animals caught and analyzed catches over many years to determine the effect of harvests on total populations and the level where exploitation would deplete the resource beyond a point of comeback.

Recent ideas of conservation in the sea have changed rather drastically. Oceanographers no longer believe that it is necessary simply to save a resource; the highest sustained yield is the goal.

Since 1950, the world fisheries have gone offshore. The Japanese were probably the greatest contributors to this change. At the end of World War II, the Japanese fishing fleet was greatly depleted, and since the nation depends so greatly on the sea —more than any other nation in the world today— rebuilding the fleet immediately to provide the nation with protein was imperative. The new fleet was built with modern ships, capable of going into any of the world's oceans in search of food.

Japanese fisheries people knew that the open ocean contained vast unexploited populations of fishes—the giant tunas, albacore, marlin, sailfish,

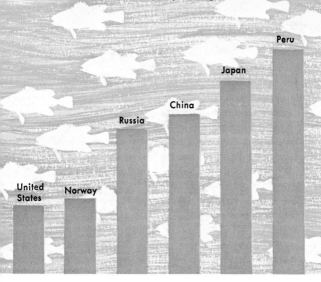

Marine harvests of the six largest producing countries in 1960's.

swordfish, and many others. With the Japanese longline or flagline, a type of fishing gear new to commercial fisheries, they swarmed over the world's oceans, fishing the deep water along the edges of the equatorial currents and in other areas where the thermocline was known to be broken. They found large, seemingly inexhaustible fisheries resources. Today Japanese vessels fish the Indian Ocean, both the North and South Pacific, and the Atlantic. Recently Soviet and American ships have joined the Japanese, and other maritime nations are also advancing their fishery methods. Large-scale cooperative fisheries investigations are being carried out now in most of the oceans of the world.

131

FOODS FROM THE SEA are almost inexhaustible. The sea contains all of the chemical ingredients necessary for life and receives over 70 percent of all of the sun's radiation striking the earth. Only the upper surface of the sea, to a depth of about 200 feet (60 meters), is in the euphotic zone where photosynthesis can occur.

The total food resources of the sea cannot be calculated solely by the bulk of organic material available, but, using the stock of fishes and squid as the primary food resource in the open ocean, the total food produced in the open ocean annually is estimated at about 2 billion tons. This figure does not include the foods harvestable from the continental shelf, bottom, and other areas.

ABOUT 19 BILLION TONS of carbon are combined annually by marine plants to form living material. In addition, the seas contain organic materials that have come from the land and chemicals released by the breakdown of previously living organisms. It is estimated that more than 90 percent of the organic matter in the oceans represents waste material from death, bacterial action, and feces rather than actual living material. This vast reservoir has been building up since the oceans were formed. In proper conditions, this material can be assimilated by plants and converted again into living organic matter.

A 90 percent loss of living material occurs at each step in land food chains. Food chains in the sea lose 60 to 75 percent at each step.

On land, photosynthesis occurs only on thin surface layer, but in the seas it occurs to a depth of as great as 1500 feet.

Europe

Africa

SARGASSO SEA

Cuba

ica

MANY SPECIES OF SEAWEEDS are harvested for food or other uses, but the only large plant found commonly in the open ocean is Sargasso Weed, a name given to several species of brown algae in the genus *Sargassum*. Some of these species drift in the open ocean, though most grow in shallow, rocky areas. In severe weather, they are torn loose in great numbers and are carried out to sea, where they collect on the surface in long windrows.

THE SARGASSO SEA, central gyre of the North Atlantic, is an area where Sargasso Weed collects in large quantities. It is never as abundant as in the myths that feature the Sargasso Sea as a place so jammed with seaweeds that sailing ships were held fast in the tangles. The seaweed is found mainly in long rows lying parallel to each other for many miles and with considerable distance of open water between. Sargasso Weed has been given much attention as a possible source of human food, livestock food, or extracts. It is rich in minerals and might yield trace elements to supplement improper diets.

Equipping large ships with nets attached to frames at the side of the vessel has been proposed to harvest seaweeds in the open ocean. These nets could be towed through rows of seaweed, and the hauls deposited in factory ship holds for immediate processing. Lack of knowledge of seasonal abundance of the weed, the great distances over which the ships must travel, and the meager market are factors inhibiting progress of projects in this particular area.

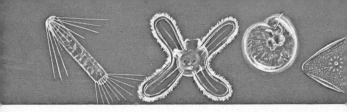

PLANKTON CULTURE may be a major means of providing food for man in the future. Direct harvest of plankton from the ocean is probably too difficult and not economical. For one thing, plankton consists of many types of plants and animals. Except in a "bloom," or great concentration of a single organism, a random ton of plankton yields at least some animals that have siliceous skeletons or bear thorny spines, as well as the stinging cells of various coelenterates and of some plants, and often kinds that are noxious. Despite these known dangers, plankton soup and plankton burgers are sometimes served to marine scientists at banquets.

Plankton abundance depends on the season and also on the nutrients available at particular times. On the whole, plankton is scattered. Some areas are practically deserts; in others there may be large blooms. Predicting where plankton will be abundant and the kind that can be harvested is not reliable enough for the establishment of an industry. Further, as in recovering chemicals (p. 132), a tremendous volume of water must be handled.

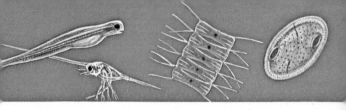

Strong currents are used to good advantage in some places to collect plankton. Off the coast of Southeast Asia, for example, giant funnel-shaped plankton nets are set in the mouths and along the courses of estuaries. A fast-flowing ebb tide carries the plankton down-river into the nets. The plankton, consisting mainly of crustaceans and small fishes, is mashed into a protein-rich paste that tastes somewhat like anchovies. It is added to rice dishes.

A variety of methods have been tried to culture phytoplankton. The most successful has been in Japan, where phytoplankton is grown in mass in outdoor ponds to which nutrients are added to cause blooms. The plankton is strained from the water, dried, and turned into a flourlike product. It is added to rice dishes, and a small portion added to breads gives them the protein equivalent of steak.

Scientists in the United States have cultured phytoplankton successfully, but the price that must be charged for the plankton flour is too high at present to attract American people from the bountiful supply of more familiar foods.

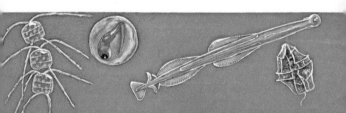

Sorting deep-sea scallops on the deck of a trawler

SHELLFISH are the shrimp, lobsters, clams, crawfish, scallops, oysters—all those animals that have some type of a hard covering. Many shellfish are taken in shallow waters, hence are not treated in this book. The shellfish fishery of the open ocean is restricted mainly to the continental shelf.

A big shellfish fishery in European waters, especially in the North Sea and adjacent waters, harvests the Norway Lobster. This small crustacean is taken by trawling on mud bottoms in several hundreds of fathoms. Great concentrations of Norway

Norway Lobster

Deep-sea Scallop

Lobsters have been located in recent years. Their tails are deep-frozen and shipped to the United States for sale in restaurants and food markets.

Off the northeastern United States and in other regions, the highly prized Deepsea Scallops, among the largest of the commercial scallops, are harvested with special dredges operated from large trawlers on the shelf. In the United States and in northern Europe, only the muscle of the scallop is eaten, and this accounts for less than half of the total flesh of the animal. Along the Mediterranean and in other parts of the world, the whole scallop is eaten. It has much the same flavor as clams and oysters.

A fishery for the Royal Pink (Royal Red) Shrimp has been developed in tropical waters. The fishery depends on two species of deep-water shrimps found at a depth of 600 to 1,200 feet (200 to 400 meters). Larger trawlers than those used for inshore shrimp fishing must be used. Because they exist in the vast, deep waters of the oceans, these large shrimp suggest an almost inexhaustible supply.

In Alaskan waters off North America, the fishery for King Crabs is large. Americans use traps; the Russians, trawls. King Crabs are found a considerable distance from shore on the continental shelf, hence have been a controversial issue among Alaskan and Soviet fishermen. To settle such problems, international treaties may be necessary, similar to those that presently exist for some other species.

Royal Red Shrimp   Alaskan King Crab

Clay pots used to catch octopus in Japan

OCTOPUSES, CUTTLEFISH, AND SQUID, known to few people, form the world's second largest potential fishery. They are eaten throughout the world, except in northern Europe, Great Britain, and among the English-speaking people of North America.

OCTOPUSES usually live in holes in rocky or coral bottoms. They are caught by hooking them with a barbless hook on the end of a short pole, by trapping them in wicker baskets, by planting long lines of pots or empty conch shells in which the animals take refuge, or by trawling. They are considered a delicacy in the Mediterranean, Orient, and Latin America. One of the largest octopus fisheries is a trawl fishery in the Saharan Bank off West Africa.

CUTTLEFISH are highly prized in the Old World, but they are not found in the Americas. Cuttlebone, collected on the beaches, is used as a dietary supplement for cage birds. It is ground to make face powders, fine molds in jewelry manufacture, and for other purposes. *Sepia*, once important as the artists' sepia, is the generic name of the cuttlefish from which the ink was obtained. One of the world's greatest fisheries for cuttlefish is off West Africa.

SQUID are harvested in the greatest quantity in Japan—about 600,000 metric tons annually of one species, *Todarodes pacificus*. Fire Squid are harvested for fertilizer. Squid are usually caught by jigging, a single fisherman catching upward of 2,000 pounds per night. The pen and ink sac are removed, and the remainder dried, usually without salt. Despite the big harvest, the demand exceeds the supply.

Fishermen jigging squid in Conception Bay, Newfoundland.

Squid are harvested commercially in United States waters only off Monterey, California, where they are taken by lampara net at night. Most of the catch is frozen and shipped abroad. The major fishery in North America is in Newfoundland, where *Illex* is caught for use as bait for cod. In 1964, a near-record year, 11,000 tons were caught.

Drum, roller, and jigs of mechanical jigger used by Japanese fishermen.

Spain and Portugal have sizable squid fisheries, but the squid are mainly canned in olive oil for exportation to other parts of Europe, Latin America, and the United States.

The supply of squid is nearly inexhaustible by present fishing methods. They are almost the sole food of the Sperm Whale and of many smaller toothed whales, porpoises, and dolphins. Most squid have a short life span. Therefore, harvesting does not over exploit the species.

SCALEFISH form the largest and most important fishery in the world at present. Important fisheries are located along all coasts, on such great fishing banks as the Georges Bank, off the New England coast, and the Grand Banks of Newfoundland, and also on the high seas. Among the most important fishing methods are hook and line, trolling, seining, trawling, harpooning, and longlining (flaglining). Electric fishing is not yet practical in the open sea.

CONTINENTAL SHELF FISHERIES are the most important in in the world. Best known of the bottom fishes are cod, hake, haddock, halibut, sole, flounder, snapper, and grouper. These fish are caught by beam or otter trawling, with bottom-fishing nets, or by hook and line. In the surface waters are the great fisheries for herring, one of the world's most important food fishes, and for menhaden, used primarily for cattle meal, fertilizer, and fish oil. These fishes are usually caught with purse seines or, in one of the English herring fisheries, with drift nets. Large trawlers capable of working greater depths permit fishing the continental slopes and tapping populations of fish that were heretofore unavailable.

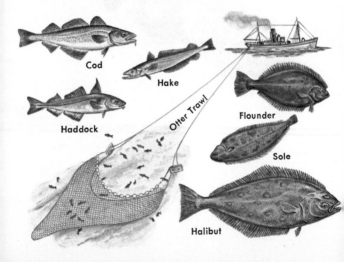

Cod

Hake

Haddock

Otter Trawl

Flounder

Sole

Halibut

HIGH SEAS FISHERIES have developed largely since World War II. Principal credit is due the Japanese and their innovation of longline (flagline) gear. With the hooks set just above the thermocline, vast catches of tuna, sailfish, marlin, and swordfish are made regularly.

Not all of the ocean is productive enough to support longlining, but along the great equatorial currents, where the thermocline is partly broken up by the turbulence of the currents, nutrients are stirred up from the depths and become utilizable. Here these great oceanic fishes congregate and have their swim paths in the oceans. Lines 50 miles (80 kilometers) long may be commercially practical even though only one hook in ten has a fish. Longlining hqs opened the oceans to exploitation.

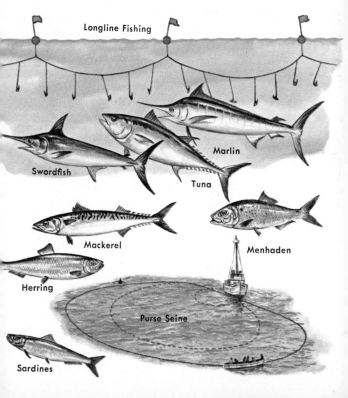

Longline Fishing

Marlin

Swordfish

Tuna

Mackerel

Menhaden

Herring

Purse Seine

Sardines

WHALING originated in the 9th and 11th centuries with the Norwegians and Basques. Later the Dutch, English and then the Americans dominated the industry. In this century the fishery developed under the influence of Norwegian techniques. By the 1952-53 season, there were 18 factory ships and 50 land stations in the whaling industry, operating 375 catcher-boats, and the catch totaled 43,669 whales. These yielded 420,000 tons of whale oil and 47,000 tons of sperm oil. During this time the fishery came under the dominance of factory ships such as the Soviet 36,000-ton *Sovietskaya Ukraina*, with over 9 catcher-boats. The major nations involved were Norway, Great Britain, the U.S.S.R., Japan, and South Africa. The fishery was centered in the pelagic whaling areas of the Southern Ocean.

As the number of whales declined, the major whaling nations united to set catch quotas in order to maintain the stocks. The idea was good, but enforcement was impossible. By the early 1960's, the whales were greatly endangered. The slaughter continued, however, and whale specialists believe that some species may have reached the point of extinction. Shore whaling stations in the Northern Hemisphere are still operating but even here the whales are endangered.

142

MARINE FISHERIES CONSERVATION differs from conservation in fresh or shallow estuarine waters. With few exceptions, fishes and invertebrates inhabiting offshore waters are not greatly influenced by the upset of natural balances by man. Some inshore fishes, such as the menhaden, are dependent in their early stages on brackish water. Chemicals dumped into an eastern river from an industrial plant a few years ago killed as many young menhaden as were caught that year by the entire Atlantic Coast fishery.

Offshore species apparently cannot be destroyed by overfishing, for when the number of individuals of a species drops to a low level, it becomes unprofitable to continue the fishery. As soon as the fishing stops, the species begins to recuperate, until fishing becomes again economically worthwhile.

POLLUTION is a constant hazard to fisheries. From wrecks, spilling of oil in ports, and dumping of bilges, an estimated 1 million tons of crude oil are dumped into the sea each year. The *Ocean Eagle* (shown here) broke up at sea off San Juan, Puerto Rico, spilling thousands of gallons of polluting oil into the water. Effective measures for coping with this problem are being studied.

Studies are being made also to learn the level of exploitation that will produce a maximum yield. Then regulations determining the amount of fishing allowed, the number of vessels, kind of gear, and other factors may be set up. Most fisheries are underfished.

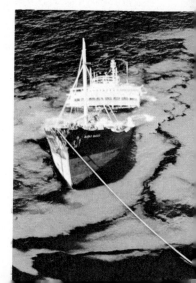

Yellowfin tuna taken aboard the *Undaunted*, a research vessel.

MANAGEMENT for conservation is complicated on the high seas because international waters can be fished by all nations. Migratory fishes are an example in which the regulations imposed by one country mean little if the next country ignores them. The Inter-American Tropical Tuna Commission was formed in an effort to manage this fishery along the west coast of the Americas. The countries involved signed a treaty permitting a study of the fishery, and each nation agreed to comply with whatever regulations are established. Similar treaties govern several important food fishes in the North Atlantic and the North Pacific. More are needed.

Atomic radiation contamination is one of the newest problems in conservation of marine fisheries. Tuna populations in the Central Pacific were contaminated and made inedible by fallout a few years ago. Radioactive cobalt and silver have been discovered recently in squid populations in Newfoundland. Only by constant vigilance can the great resources of the sea be kept for future generations.

FISHERIES OF THE FUTURE have great potential. Based on present fishery methods and marketing, the fisheries cannot be exploited beyond definite levels. Yields can be increased only by: (1) more complete utilization of the present catch; (2) discovery and exploitation of new resources; (3) increasing productivity; or (4) intensified farming of the sea.

MORE COMPLETE UTILIZATION of fishery resources is needed. On the Tortugas shrimp grounds, more than half of each trawl haul consists of bottom fish other than shrimp. All of these trash fish—that is, kinds of fish other than those wanted—are thrown overboard. In the future, these fish may be utilized in making a highly concentrated "fish flour."

Marine Protein Concentrate (MPC), a nearly tasteless, odorless flour produced by several different methods of processing marine organisms, provides the full protein requirements for both livestock and humans. A recent United States survey reported that MPC, consisting of 85 percent protein, can be produced for 25 cents per pound or less. As an additive, a pinch of MPC in one biscuit can furnish the equivalent of the protein in a pound of beef. It may be added to bread, rice dishes, and other foods. For human consumption, the processing is extensive, and full precautions are used to assure a safe, healthful product.

Protein Concentrate processing plant, New Bedford, Mass.

MPC may truly be the food resource solution for a starving world. Livestock have been fed a far less purified form for many years.

Conventional fisheries may not be able to raise the fishery yields to much more than double the present 10 percent of the world's food needs. But by processing the presently unused bottom organisms of the sea to make MPC, the oceans' full productivity, estimated at about 400 billion tons wet weight per year, can be more fully used.

DISCOVERIES OF NEW FISH-ERIES RESOURCES are always significant. Many, such as the Royal Red Shrimp fisheries, the offshore scallop fisheries, and the development of the oceanic tuna fisheries, have been spectacularly successful. But mainly these have been discoveries of new populations of an already utilized species. The end of their commercial numbers is foreseeable.

The depths of the sea—that vast area lying below the photic zone—are hardly touched. A small fish, *Cyclothone*, that lives in these deep waters is probably the most common fish in the sea. Hatchetfishes, likewise cosmopolitan, are also abundant. Tremendous schools of deep-sea squid, numerous kinds of deep-sea bottom-dwelling crabs and lobsterlike crustaceans, and large brotulid fishes are among the others yet unexploited. New kinds of fishing gear and different ship design will be needed for some of these harvests that represent a valuable food resource. Many will probably be processed to make protein-rich "fish flour" (p. 145).

THE HARVEST POTENTIAL from the sea depends on the food produced by plants, which in turn is determined by availability of sunlight and nutrients. Since the sunlight is a fixed quantity, the variable productivity of the sea is determined by the utilizable amounts of phosphates, nitrates, and other elements that are present. By increasing the amounts of these "fertilizers" in a fishing area, the yield of fish may also be increased.

The challenge is to lift nutrient-rich deep waters to the sunlight in offshore locations. The use of thermal pumps to circulate the water from the depths has been proposed, as has the construction of dams in the path of deep currents to shunt the water upward. Some oceanographers have suggested that controlled deep-sea atomic explosions in specified areas might cause permanent upwellings and form new centers of productivity. Diverting the course of some ocean currents toward more favorable areas has been proposed, too. Methods such as these may be carried out with dramatic results in the future, but we need much more careful study before we dare upset nature's balance.

FARMING THE SEA represents one of the greatest potentials in future fisheries. Man changed from a hunter to a husbandman on land, but in the sea he is still largely a hunter. There are a few exceptions, such as oyster culture as practiced in France, Japan, and to a lesser but increasing degree in the United States.

To farm the sea, man must go beneath the surface. Undersea farms, probably leased from the government, may be plowed by submarine tractors

Aerial photograph of culture beds of *Porphyra* (alga) in Tokyo Bay.

to release nutrients buried in the silts and to increase productivity. Other innovations are foreseeable. Undersea farms will be raked or swept to get rid of all the unproductive but demanding branches in the food chain—such as starfish, fish too small to be utilized, or trash fish that might utilize food required by the wanted fish or scallops. The harvesting will be done by submersibles towing nets rather than by bottom trawling from surface vessels. The catch will be detected and then followed either by television or by direct observation.

Sea routes followed by migrating fishes may some day be fully plotted. Then the fishes may be diverted from their usual feeding grounds to ones more productive in a particular year. By sounding devices, whales may some day be herded seasonally to richer pastures, and the annual crop of calves thinned in a systematic fashion, as with cattle. Only surplus adults and those that are past their productive prime will then be utilized in the fishery.

The first great steps in sea farming will take place inshore, where control of the environment is not as difficult as it is in the open sea. But it is certain that future generations will see the oceans utilized as fully and efficiently as we presently utilize the land.

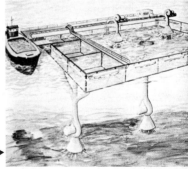

An artist's concept of an open-sea farming operation.

147

This comparatively new field in the closely related ocean sciences evolved as entirely new approaches in observational methods and data collecting were developed with modern oceanography. Utilization of the sea's vast resources requires new techniques for harvesting—from the surface to the very bottom of the ocean basins. Beach erosion, undersea habitats, bulkheading, undersea instrument installation, and acoustical studies all fall within the domain of the ocean engineers. Government agencies, universities, and private industries all have need for properly trained and qualified ocean engineers. As a career, this new field offers almost unlimited opportunities and exciting challenges.

OCEAN RECORDING SYSTEMS provide a new technique for obtaining data from the sea for basic oceanographic research, submarine warfare, navigation, and weather forecasting.

In the past, oceanographers have had to rely upon series of observations taken by single or multiple-ship operations. For reliable computer analysis, however, data should be collected over long periods from many areas at the same time. The information could be collected from ships, but many ships would be needed. They are expensive, require much manpower, and cannot be anchored for the long periods necessary. For this reason, ocean engineers have developed giant buoys to gather information offshore. These monster buoys contain great arrays of instruments for recording data about salinity, temperature, currents, wave force, solar radiation, cloud cover, wind force, and many other factors. The information is either stored in the buoy system or relayed by radio to shore stations at regular intervals.

On the surface, the giant buoys are both a danger to ships and are endangered by them. Storms may also cause loss or damage. For this reason, buoy systems have been devised that can be anchored to the bottom. They gather information at various levels by unwinching the cable, then rise to the surface and transmit their data to ship or shore station; then they haul themselves back to the bottom again. These entirely automated, widely dispersed, and carefully located deep buoys offer almost unlimited possibilities in ocean research. They will provide a wealth of information that could not be collected otherwise without whole fleets of surface vessels.

Biological systems of this type are yet to be developed. Oceanic biology continues to lag in this new phase of oceanography because the problems of surveying organisms and recording their sizes, numbers and possibly even different kinds are much more involved.

BUOYS are much less costly to keep on station than are ships. This monster buoy (shown at right) was anchored in the sea about 1,000 miles north of Hawaii by scientists from Scripps Institution of Oceanography. The program, sponsored by the Office of Naval Research, uses meteorological and oceanographic data sent back daily from this giant untended buoy and from several other similar stations.

COASTAL ENGINEERING deals largely with the ravages of the sea upon the shores. Erosion and the washing away of shores has plagued man since earliest times. But along such areas as the sandy beaches from Long Island to Florida, the windswept coasts of Britain, and the low, dyked land of Holland, man's battle with the sea is never-ending. Engineers must learn how to stop direct erosion, how to stabilize beaches by constructing groins and offshore barriers, and how to build up beaches again by dredging. Filled-in estuaries, crowded with housing developments, impede the flow of water. Engineers must work out ways to increase the flow and thus avoid overpollution. Engineers can design porous bulkheads; they can also reduce erosion by shaping the bulkheads so that they weaken wave impact and at the same time eliminate cross waves. Harbors and jetties, channel structures, offshore sewage disposal units, artificial reefs —these are among the projects that need the know-how and experience of coastal engineers.

Developments have destroyed valuable estuaries in Tampa Bay area.

OFFSHORE OIL PLATFORMS are so numerous along parts of the Gulf of Mexico that their lights at night resemble those of a small city. The continued search for oil will result in many more of these being built along the shores of the world.

Despite utmost care and the installation of the latest safety devices, leakage and oil spills occur, polluting the inshore beaches. On the other hand, in some areas the platforms provide shelter for sport fish that seek them out to feed on the accumulated life growing on the steel and concrete pilings.

OCEAN DRILLING AND MINING involve engineering skills, many of which are already highly developed in the oil and gas industries. The first offshore oil well was constructed in 1948 off Louisiana. Now offshore oil rigs and platforms, often several miles at sea on the edge of the continental shelf, dot shores from the Gulf of Mexico to New England and from the Gulf of Guinea to the North Sea. With their widespread use has come the problem of oil leakage and spills. Prevention of oil pollution in the sea is a major problem.

The sea's greatest resource is its more than 1 billion cubic kilometers of water, enough to supply all of the water needed by the largest imaginable concentration of human life. But sea water contains in solution almost all known chemicals and must be purified before it can be used.

151

Giant desalination plant supplying fresh water for San Diego, Calif.

THE COST OF HANDLING the great volume of water required to supply human needs is the greatest problem, but new methods are sharply reducing these costs. Fresh water from the sea is now being supplied to coastal cities in several areas of the world, such as Key West and Curacao.

Bromine and magnesium are important extracts from sea water. The shallow waters on the continental shelf yield sizable mineral deposits that

MOST ABUNDANT COMPOUNDS AND ELEMENTS IN ONE CUBIC MILE OF SEAWATER

	TONS		TONS
Sodium chloride	128,284,403	Magnesium chloride	17,946,522
Magnesium sufate	7,816,053	Calcium sulfate	5,934,747
Potassium sulfate	4,068,255	Calcium carbonate	578,832
Rubidium	64,316	Fluorine	1,400
Barium	916	Zinc	450
Iodine	90	Arsenic	to 368
Phosphorus	to 400	Nitrogen	to 1,300

From *Marine Products of Commerce*, by Donald K. Tressler and James McW. Lemon, Van Nostrand Reinhold Publishing Corp., New York, N.Y., 1951.

can be mined. Potential concentrations of gold, platinum, and chromite are found along the Pacific Coast of the U.S.; phosphorite, ilmenite, and zircon occur along the Atlantic Coast. Nodules, crusts, and oozes in the deep sea contain significant amounts of manganese, copper, cobalt, and nickel. Diatomaceous oozes, almost pure silica or glass, form the diatomaceous earth used in many industries. Deep diatomaceous oozes found over much of the ocean floor may some day replace the diminishing supplies on land. Magnetite is mined off Japan; tin off Malaysia; thorium sands off South India. Diamonds are taken by dredges off South Africa.

With dwindling supplies of minerals on land, ocean engineers are designing new mining techniques to tap these rich resources. The Law of the Sea is an important factor regarding the ownership and right of exploitation of resources on the continental shelf and in the deep sea. These and other problems need solution in this field of ocean engineering, one of the newest and fastest-growing branches of oceanography.

MANGANESE NODULES are found over large areas of the ocean floor in quantities estimated at 30,000 to 55,000 tons per square mile. They may contain as much as 50 percent pure manganese, varying from ocean to ocean. They are spread widely, and so collecting these nodules for commercial purposes requires a special dredge or trawl. The nodules in the picture were photographed at a depth of about 10,000 ft. in Drake Passage off the southern tip of South America.

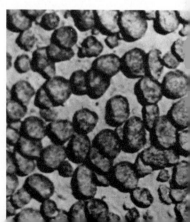

LIFE SUPPORT SYSTEMS in which man can explore and work beneath the sea are a major triumph of ocean engineering.

Man's entrance into the sea was a slow process, progressing over the years from shallow free diving without aids to suit diving, and finally to the Aqualung, developed in 1942 by Cousteau and Gagnon, at last setting man free in the shallow sea. In the deep sea, similarly, man advanced from the early tethered diving vessel of Beebe to Cousteau's diving saucer and Piccard's bathyscaphe, and finally to a vast array of deep-diving vehicles (DDVs), or submersibles, developed by industry. Today, man needs no longer to depend on trawls and grabs to explore the depths; he can descend into the depths himself to see what is there and to bring back their treasures.

Each year submersibles with greater maneuverability, greater depth capabilities, and more facile collecting devices are produced. The vessels are equipped with giant viewing ports, searchlights to probe the unlighted depths, and legs for resting on the bottom. With wheels, they will be able to roll along the ocean floor. Extending from below the viewing ports are various types of mechanical arms capable of collecting rock samples, directing drill bits into their casings, recovering instruments from the sea floor, and performing many other feats as man's arms in the deep sea.

With the ever-increasing versatility and operational depths of DDVs—8,310 feet in 1968—the mysteries of the deep sea are being unveiled. Ocean engineering is directed toward a future in which man will reign supreme in the sea.

Tektite I, for underwater living and research in Virgin Islands

MANNED UNDERSEA HABITATS are the most recent products of science and technology in their attempts to probe the sea. We are still far from the realization of undersea cities and tourist resorts, but there are needs now for safe undersea habitats in which scientists and engineers can live and work for long periods of time. Here they can conduct studies of the undersea life environment and learn also about the stresses of such living conditions on humans. Undersea living quarters may eventually be bases for mining operations, for testing of oceanographic equipment, for operation of

155

defense monitoring systems, and perhaps for submerged atomic-power generating plants.

Engineers and scientists are concerned today with submerged manned stations in only a few hundred feet of water. In the future, manned stations thousands of feet deep will be needed. They will require large support DDVs to transfer personnel and supplies and many smaller, highly maneuverable scout and work vessels. These wide-ranging, smaller vessels will keep in constant contact with their bases by means of refined acoustical communication systems.

Man's equipment for exploration of the depths will soon make Jules Verne's *Nautilus* as obsolete as early television and the first orbiting satellites.

A Conversion Table of Common Measurements

1 inch = 25.4 mm. or 2.54 cm.

1 foot = 0.3048 meters

1 yard = 0.914 meters

1 nautical mile = 1.852 kilometers

1 statute mile = 1.609 kilometers

1 mm. = 0.039 inches

1 cm. = 0.3937 inches

1 m. = 39.37 inches = 3.28 feet

1 kilometer = 0.621 miles

1 lb. = 0.453 kilograms

1 kilogram = 2.2 pounds

1 fathom = 1.83 meters = 6 feet

1 knot = 1.15 statute miles per hour = 1 nautical mile = 1.852 kilometers per hour

1 statute mile = 5,280 feet = 0.868 nautical miles

1 nautical mile = 6,076 feet = 1 minute of the meridian at the place of observation

for practical purposes, 1 nautical mile = 1,000 fathoms = 2,000 yards = 1" of longitude

to find the number of feet per number of meters, multiply the number of meters by 3.28

to find the number of meters per number of feet, multiply the number of feet by .3048

to find the number of meters per number of fathoms, multiply the number of fathoms by 1.83

Note: The conversion figures given in the text in this book are rounded to avoid indication of a preciseness that did not exist in the original figures.

MORE INFORMATION

Bardach, John, *Harvest of the Sea.* George Allen and Unwin, London, 1969.

Carson, Rachel, *The Sea Around Us.* Oxford University Press, New York, N.Y. Rev. ed., 1961. The classic popular book on the oceans.

Cousteau, J. Y., and J. Dugan, *The Living Sea.* Harper and Row, N.Y., N.Y. An account of life in the sea.

Ericson, David, and Goesta Wollin, *The Ever-changing Sea.* Alfred A. Knopf, N.Y., N.Y., 1967. A popular account of recent oceanographic research and discoveries.

Friedrich, H., *Marine Biology.* University of Washington Press, Seattle, Wash., 1970. A general account for students and the non-marine biologist. Primarily deals with the open sea.

Gross, Grant M., *Oceanography: A Review of The Earth.* Prentice-Hall, Englewood, N.J., 1972. Introductory oceanography with emphasis on shores.

Groven, P., *The Waters of The Sea.* Van Nostrand, London, 1967. Readable account of waves, currents, tides, and ice formation.

Hardy, Sir Alister C., *The Open Sea: Its Natural History.* Vol. 2: *Fish and Fisheries.* Houghton Mifflin, Boston, Mass., 1959. A very readable popular work on fishes and fisheries of the North Atlantic.

Harvey, H. W., *The Chemistry and Fertility of Sea Water.* Harvard University Press, Cambridge, Mass., 1960. Introduction to chemical oceanography.

Herring, Peter J., and Malcolm B. Clarke, *Deep Oceans.* Praeger Publishers, N.Y., N.Y., 1971. Authoritative, interesting, and well-illustrated.

King, Cuchulaine A., *An Introduction to Oceanography.* McGraw-Hill Book Co., New York, N.Y., 1963. A general introduction to physical oceanography and marine geology. College level.

Marshall, N. B., *Aspects of Deep-sea Biology.* Hutchinson, London, 1958. Life in the deep sea and its adaptations.

Murray, J., and J. Hjort, *The Depths of The Oceans.* Macmillan and Co., Ltd., London, 1912. Still the best book on oceanography. Facsimile reprint but expensive.

"Sea Frontiers," published by the International Oceanographic Foundation, Miami, Florida. A popular journal devoted to all aspects of the oceans. Readable, authoritative.

Shenton, E. H., *Diving for Science.* W. W. Norton, N.Y., N.Y., 1972. Story of the development of deep submergence research vessels.

Sverdrup, H. U., M. W. Johnson, and R. H. Fleming, *The Oceans: Their Physics, Chemistry, and General Biology.* Prentice-Hall, Inc., New York, N.Y., 1942. The oceanographer's "bible," a comprehensive detailed reference for the professional oceanographer.

Weyl, Peter K. *Oceanography: An Introduction to The Marine Environment.* John Wiley, N.Y., N.Y., 1970. A physical view of the oceans.

INDEX

159

C   D   E   F   G